SpringerBriefs in Applied Sciences and Technology

SpringerBriefs present concise summaries of cutting-edge research and practical applications across a wide spectrum of fields. Featuring compact volumes of 50–125 pages, the series covers a range of content from professional to academic. Typical publications can be:

- A timely report of state-of-the art methods
- An introduction to or a manual for the application of mathematical or computer techniques
- A bridge between new research results, as published in journal articles
- A snapshot of a hot or emerging topic
- An in-depth case study
- A presentation of core concepts that students must understand in order to make independent contributions

SpringerBriefs are characterized by fast, global electronic dissemination, standard publishing contracts, standardized manuscript preparation and formatting guidelines, and expedited production schedules.

On the one hand, **SpringerBriefs in Applied Sciences and Technology** are devoted to the publication of fundamentals and applications within the different classical engineering disciplines as well as in interdisciplinary fields that recently emerged between these areas. On the other hand, as the boundary separating fundamental research and applied technology is more and more dissolving, this series is particularly open to trans-disciplinary topics between fundamental science and engineering.

Indexed by EI-Compendex, SCOPUS and Springerlink.

More information about this series at http://www.springer.com/series/8884

K. Kunaifi · A. J. Veldhuis ·
A. H. M. E. Reinders

The Electricity Grid in Indonesia

The Experiences of End-Users and Their
Attitudes Toward Solar Photovoltaics

 Springer

K. Kunaifi
Department of Design, Production
and Management
Faculty of Engineering Technology
University of Twente
Enschede, The Netherlands

Department of Electrical Engineering
Faculty of Science and Technology
UIN Suska Riau University
Pekanbaru, Indonesia

A. H. M. E. Reinders
Department of Design, Production
and Management
Faculty of Engineering Technology
University of Twente
Enschede, The Netherlands

Department of Mechanical Engineering
Energy Technology Group
Eindhoven University of Technology
Eindhoven, The Netherlands

A. J. Veldhuis
Department of Design, Production
and Management
Faculty of Engineering Technology
University of Twente
Enschede, The Netherlands

Alliander N.V.
Arnhem, The Netherlands

ISSN 2191-530X   ISSN 2191-5318   (electronic)
SpringerBriefs in Applied Sciences and Technology
ISBN 978-3-030-38341-1   ISBN 978-3-030-38342-8   (eBook)
https://doi.org/10.1007/978-3-030-38342-8

This Springer imprint is published by the registered company Springer Nature Switzerland AG
The registered company address is: Gewerbestrasse 11, 6330 Cham, Switzerland

Preface

In 2017, more than 60 million households in Indonesia were connected to the national power grid. Accordingly, we believe that their 'voice' is important to maintain democratic and participatory values in planning electricity services. These voices should be taken into account by the utility company and other policymakers. However, what is actually the voice of electricity users in Indonesia? Also, what can we learn from it when looking at the fitness of the electricity supply in Indonesia in the context of costs, the reliability, and environmental aspects?

This book presents the real experience of households, some of the grid users in Indonesia. Through a series of surveys in 2017, households in three cities in Western, Central, and Eastern Indonesia shared their experiences and preferences regarding their electricity supply. They offered their opinions about the stability and reliability of electricity supply, how they coped with blackouts, and what impacts power interruptions had on their daily lives.

Because of the frequent power outages, the users started to think about the importance of having a back-up power generator at home. Given that Indonesia has high solar irradiance the whole year through, we also observed the users' attitudes toward solar photovoltaic (PV) systems.

The book starts with a brief introduction about Indonesia followed by the status and challenges of power supply in the country. Then, in the middle section, the users' voices are presented. Finally, the potential of PV systems, as a promising solution to increasing electricity access and improving the reliability of electricity supply in this tropical country, is presented. We believe that this book provides useful information for the transition to the use of solar energy in energy systems in Indonesia, which is meant for academia, electric utility companies, PV system actors, policymakers, and of course, households in Indonesia.

Enschede, The Netherlands   K. Kunaifi
Arnhem, The Netherlands   A. J. Veldhuis
Enschede/Eindhoven, The Netherlands   A. H. M. E. Reinders

Contents

Chapter 1
Energy in Indonesia: The Main Factors

1.1 Preface

This chapter will introduce the energy situation in Indonesia as well as the main factors that influence it. Being one of the world's largest archipelagos, Indonesia has a unique and highly distributed power supply system. The population size is the main factor which causes a significant demand for energy. The growing economy of Indonesia brings optimism, including on the subject of renewable energy development. Across Indonesia's area, high levels of solar energy are available the whole year through. However, the present role of the central government in energy development is still very big, which causes inefficiency. This demands a more significant role of local entities, especially governments and the private sector. In our discussions, we mainly refer to statistics from 2017 in combination with older data.

1.2 An Archipelago with a Unique Power Supply System

Indonesia is one of the world's largest archipelagos. The country has a total area of 8.3 million km^2, comprising 77% water surface and 17,504 islands [1]. Around 111 of its outer islands share borders with neighboring countries. As shown in Fig. 1.1, Indonesia stretches over 5,000 km from the East to the West, across South-East Asia and Oceania and two oceans, the Indian Ocean and the Pacific. Most of the citizens live on five major islands, namely Sumatra, Java, Kalimantan, Sulawesi, and Papua. Yet, around 6,000 other islands were inhabited in 2014 [2].

© The Author(s), under exclusive license to Springer Nature Switzerland AG 2020
K. Kunaifi et al., *The Electricity Grid in Indonesia*,
SpringerBriefs in Applied Sciences and Technology,
https://doi.org/10.1007/978-3-030-38342-8_1

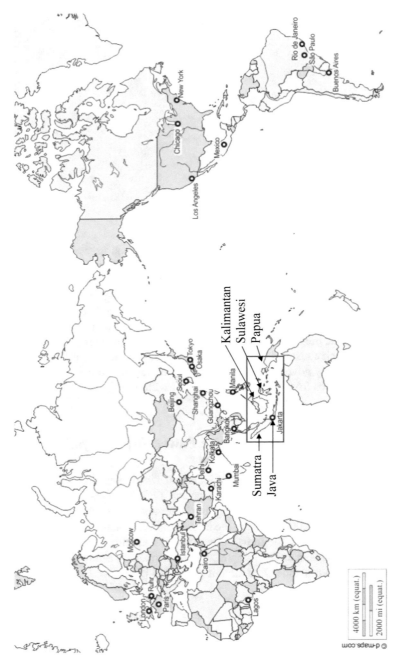

Fig. 1.1 The position of Indonesia in the world map as shown in the red box [49]

The land area of Indonesia covers 42% of the entire land of South-East Asia,[1] with dimensions that are larger than the total area of the European Union (EU) member nations.

Two of the Indonesia's islands, Kalimantan and Sumatra, are among the top ten largest islands on Earth. Kalimantan or Borneo is the fourth largest island in the world, with a size of 539.460 km^2. Sumatra is the seventh largest island, with size of 473.606 km^2. Indonesia's territory on the island of New Guinea, the second largest island in the world, is roughly equal to half of the island's total area. With so many large islands, the length of the coastline of Indonesia is 108,000 km [1].

Geologically, the Indonesian archipelago has a complex structure [3]. On the island of Sulawesi in the center of Indonesia, the meeting point of three large tectonic plates is located, namely Eurasia, the Pacific, and Australia. Accordingly, the majority of Indonesia is also part of the Pacific Ring of Fire, a volcanic path along the Pacific Ocean. Such a geological attribute gives Indonesia a great potential for geothermal energy. From around 312 sites in Java, Sumatra, Bali, Nusa Tenggara, and Sulawesi, Indonesia has approximately 28.9 gigawatt (GW) of geothermal energy potential [4].

In line with its geographical nature, Indonesia is quite familiar with many disasters, both natural and anthropogenic. The main types of disasters are floods, droughts, earthquakes, volcanic eruptions, tsunamis, and forest fires [5]. Seventy-six active volcanoes, mostly on Java and Sumatra, contribute to frequent volcanic eruptions and earthquakes. As a very active seismic zone, Indonesia is not only the country with the most earthquakes events in the world but also with the highest number of earthquakes per unit area, after Tonga and Fiji [6].

Most of Indonesia's topography consists of lowland and coastal areas, as well as mountains on the large islands [5]. Around 5,700 rivers flow across the country, which contributes to about 12% of its land being suitable for farming [7]. Fertile soil, mainly originating from volcanic ashes, is mostly located on the islands of Java, Sumatra, Sulawesi, and Nusa Tenggara. Half of Indonesia's land is currently covered by rainforest, mainly located in Sumatra, Kalimantan, Papua, and Sulawesi. However, around 30% of the natural forest has been converted to other uses, such as agriculture, mining, infrastructure, and urbanization [8]. Some locations, such as the eastern coast of Sumatra, the southern coast of Kalimantan, and the northern coast of Java, are dominated by swamps and mangrove forests.

Given the characteristics described above, especially due to its geography as an archipelagic country and the size of the area itself, Indonesia faces unique challenges as well as opportunities concerning electricity services. The most common challenge is due to the existence of islands, which causes difficulty in providing evenly distributed electricity service to all the regions of Indonesia. This also brings a unique opportunity to expand the potential of its renewable resources for power generation.

---

[1] South-East Asia consists of 11 countries, namely Brunei, Cambodia, Timor-Leste, Indonesia, Laos, Malaysia, Myanmar, Philippines, Singapore, Thailand, and Vietnam.

The distribution of access to electricity[2] service in Indonesia is not balanced between regions. On islands with high industrial intensity, such as western Java [11], the quality of electricity access is very good. Also, in the capital city of Jakarta, the electrification ratio, ER, is almost 100%. However, in other areas, particularly in the eastern part of Indonesia, access to electricity is relatively low. The ER in 2017 in the Province of Papua, for example, was only 39% [12].

People who live on smaller islands far away from the main islands, where the main grid exists, cannot conventionally obtain access to electricity from the main utility networks. Either from a technical or economic perspective, it is often not feasible to extend the transmission networks to remote islands [13].

Constructing power plants on remote islands requires great cost and high levels of complication [14]. Therefore, for decades, diesel generators have been very popular on islands in Indonesia due to their simple installation and low initial costs. A 200 kilowatt (kW) diesel generator, for example, requires an initial investment of around US$ 200,000.

However, in reality, diesel power generators are not favorable because of their critical weaknesses such as noise pollution, soil and water pollution, insecure diesel fuel supply, skyrocketing diesel fuel prices, and greenhouse gas (GHG) emissions from burning diesel fuel. In the long term, the use of diesel generators on islands will not be sustainable. Therefore, the use of locally available renewable energy resources for power generation is the future of power systems on Indonesia's islands. Solar photovoltaic (PV) systems, in particular, are a rational option for the development of the electricity sector in Indonesia's archipelago.

Photovoltaic systems are among the most promising energy technologies of the twenty-first century. Three obvious reasons support this assertion [15]. First, PV systems are a clean way to generate electrical energy because their operation does not involve any GHG emission nor pollutants that can cause global warming and acid rain. Secondly, PV systems convert solar irradiance, which is always available, into electricity. Finally, solar energy is available in abundance: the amount of solar irradiation that reaches the Earth in 1 h is equivalent to the world's total energy consumption in one year.

Indonesia has high levels of solar energy that are available the whole year through. However, it is important to ask: Is Indonesia ready to adopt solar electricity or not? By understanding Indonesia's overall characteristics such as geography, population, economy, climate, and stakeholders in the energy sector, we have provided an overview of how solar electricity fits with Indonesia. This can create insight into the opportunity and drawbacks of solar electricity in Indonesia.

---

[2] Access to electricity can be represented by the electrification ratio (ER) or electrification level (EL), which is defined as the percentage of population or households with electricity. The World Bank, IEA, UN metadata, and other international organizations, however, use the term 'electrification rate' instead, although in some literature, the electrification rate is also defined as the price of electrical energy [9, 10]. In this book, we use the term electrification ratio to quantify access to electricity.

1.3 A Growing Economy that Brings Optimism

At present, the Indonesian government is working hard to spur social development and economic growth by improving the economy, eradicating corruption, cutting domestic fuel subsidies, and boosting investments in infrastructure [7]. In 2018, Indonesia's economy grew by 5.2%. Although this figure was high compared to the average economic growth of the world of around 2% [16], Indonesia's government targets a growth rate of 7% by stimulating foreign investment [7]. In South-East Asia, Indonesia surpassed Malaysia and Thailand in terms of economic growth, but was below the Philippines and Vietnam. Higher economic growth rates are common in emerging economies, given the examples of India, Bangladesh, and Ghana, with greater economic growth in the range of 7–8%. In contrast, the developed countries like the United States (US), Singapore, and Australia stayed in the range of 2% growth.

Regarding the gross domestic product (GDP) based on purchasing power parity (PPP), Indonesia is predicted to become one of the world's economic leaders in the coming decades [17]. In 2018, Indonesia's GDP (PPP) was US$ 3.5 billion, which was ranked seventh in the world, [18] and shared about 2.6% of the total global GDP (PPP). However, due to its large population, the GDP (PPP) per capita was low (around US$ 13,057) or in the 86th place globally. To compare, the GDP (PPP) per capita of Malaysia, the US, and Singapore were US$ 31,698, US$ 62,641, and US$ 101,353, respectively [18].

There is a strong correlation between the GDP of a country and its energy consumption [19, 20]. Figure 1.2 shows the per capita energy consumption (in kg oil equivalent) versus per capita GDP, PPP (current international $). The size of the bubbles denotes the total population per country. All values, however, refer to the year 2014, when the population of Indonesia was 255 million [21], GDP per capita (PPP) was US$ 10,570 [18], and energy consumption per capita was 884 kg oil equivalent [22]. As shown by the blue fitting line, countries with a higher GDP per capita used more energy than those with lower GDP per capita. In terms of GDP (PPP) per capita, Indonesia was among the countries with upper-middle income. Also, energy use per capita in Indonesia was less than half of the world's average. The economic size and the current level of energy use per capita indicate that a sharp increase in energy use can be expected in the coming years.

In 2017, four sectors contributed most to Indonesia's GDP. They include the processing industry, which contributed to 20.2% of total GDP, followed by agriculture, forestry, and fisheries at 13.1% [23]. Retail and trade, car and motorcycle repairs are ranked at the third position with 13% followed by construction at 7.8%.

Regarding inflation, during the past five years, Indonesia's rate was 4.2%, which was categorized as healthy [24]. The inflation rates in some member countries of the Association of South-East Asian Nations (ASEAN) were 6.8% in Myanmar, 3.8% in the Philippines, and 3.5% in Vietnam. Brunei Darussalam and Singapore had the smallest inflation rates of 0.2 and 0.7% [25].

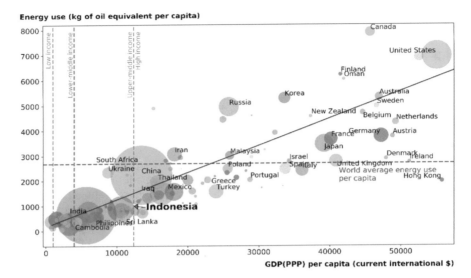

Energy use (kg of oil equivalent per capita)

Fig. 1.2 Correlation of per capita energy consumption (in kg oil equivalent) and per capita GDP, PPP (current international $). The graph is plotted based on data in 2014 from [18, 21, 22]

Does economic condition relate to the development of renewable energy in a country? The Environmental Kuznets Curve (EKC), a well-known curve in energy and sustainability disciplines proposed by Kuznet (1955) (Fig. 1.3), can be helpful to indirectly answer this question. The curve shows the relationship between the economic growth of a country and environmental quality. According to the EKC hypothesis, shown as a horseshoe curve, in the early stages of per capita income growth, environmental quality decreases. But, after some level of income per capita, the tendency reverses, so that the environmental improvement follows high-income levels [26]. Based on this concept, the relative level of environmental degradation of countries can be mapped based on information about their economies [27].

Fig. 1.3 Environmental
Kuznets Curve (EKC) [50]

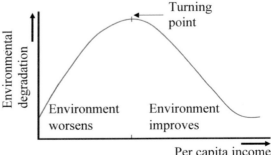

The relationship between the economy and renewable energy can be suggestive given that the primary motivation to boost renewable energy worldwide is to tackle the apparent degradation in the global environment, particularly the issues of global warming and climate change. The alarming climate condition of planet Earth due to the exploitation of natural resources for energy generation can be considered as already reaching the top of the curve, and therefore attempts to reduce it globally can be referred to the turning point of the curve [28]. After that, the use of renewable energy should be kept stable or increased [20] along with stable growth in the economy.

The growth in renewable energy, especially solar and wind power, is an essential factor in ensuring sustainable economic growth, contributing significantly to improving the socio-economic condition of a country [29]. Economic opportunities exist at each phase of the energy business chain. Starting from project planning, manufacturing processes, installation, network construction, maintenance processes, to the end of the project, all have positive economic and employment impacts.

The economic strength of a country illustrates its ability to develop new sectors such as renewable energy. However, economic power is not the sole success factor for renewable energy development. GDP per capita is also important for renewable energy technologies because it determines the purchasing power of individuals. Finally, the state policy is also very decisive, whether it is positive, contradictory, or neutral in the war against environmental damage, especially climate change.

1.4   Big Population with a Significant Demand for Energy

Indonesia is currently the fourth most populated country in the world after the US, India, and China [30]. The total population in 2017 was 261 million [23]. Representing 40% of the total population of South-East Asian countries, Indonesia has a vital role in shaping the population distribution of the ASEAN [31]. The population distribution, however, varies widely from one island to the other. Around 57% of the population inhabits the island of Java alone, totaling 149 million people, while 20, 7, and 6% live on the islands of Sumatra, Sulawesi, and Kalimantan, respectively. Ten percent of the population live in the rest of Indonesia, with a total area of 2,300,000 km^2. As such, Indonesia has large differences in population density.

In the province of West Java alone, around 48 million people live, which exceeds the total population of the islands of Kalimantan, Sulawesi, Papua, Maluku Islands, and ENT combined (about 47.5 million people). East Java follows as the second most populous province with 39 million, and Central Java thereafter with 34 million people. Jones [31] predicted that by 2035, Indonesia would have an additional 67 million people compared to the population in 2010, of which Java would contribute 30 million.

Although most of the population occupies the five major islands, a significant number of people live on smaller islands that are mainly located in the eastern part of Indonesia. In 2016, nearly 13,000 (16%) of 82,030 villages in Indonesia were categorized as undeveloped because they were rural villages or settlements on isolated islands [32] with a lack of access to electricity. Figure 1.4 shows the population per province in millions of people.

Although the projected increase in the population looks massive, the rate of population growth in the period 2010–2035 is estimated to decrease [31]. At present, the national rate of population growth is 1.4%. By 2035, the population growth would be around 0.6%, which is far less than the current level. Also, in line with the lower level of population growth in Java and migration between regions, Java is projected to experience a 2% population decline in 2035.

The province of Jakarta has a very dense population compared to other provinces, with 15,624 people per km^2, while West Java has 1,358 people per km^2 [33]. However, when compared with the most populated cities in the world, the population density in Jakarta is far below Manila with 41,515 people per km^2, and Mumbai with 28,508 people per km^2 [27]. The provinces of North Kalimantan and West Papua have the lowest population density in Indonesia, with only 9 people per km^2. The national average population density is 137 people per km^2. To compare, Brazil, the fifth most populous country, has a population density of only 25 people per km^2 [34].

1.5   A Challenging Climate with Excellent Solar Energy Potential

Solar PV systems are part of a rapidly growing renewable energy technology which is increasingly playing an important role in reducing dependence on conventional fossil fuels for electricity generation. A PV system converts sunlight into direct current (DC) electricity by using semiconductor solar cells wired to each other in PV modules. Multiple modules can be connected to form an array, which can be scaled up or down to produce the desired amount of power. Nowadays, commercial PV modules such as silicon modules convert around 17% of the incoming solar irradiance [35] and can last for 25 years, producing sustainable electricity across the globe.

The total global installed capacity of PV systems by the end of 2018 was around 515 GW; 95% was crystalline silicon, with around 443 Terawatt-hour (TWh) electricity production in 2017 [35]. Although the most commonly used materials today are from the silicon family, other materials are being tested and used. These include gallium arsenide (GaAs), hybrids, chalcogenides (e.g., cadmium telluride/CdTe and copper indium gallium selenide/CIGS), and other emerging photovoltaics (e.g., organic and perovskite).

A typical PV system comprises an array made of solar panels, an inverter, and other electrical hardware also called BOS. The two principal classifications of PV

Fig. 1.4 Map of the total population by province (in millions of people) [23]

systems are stand-alone and grid-connected systems. A stand-alone PV system operates independent of a public grid and is generally sized and designed to meet certain AC or DC loads.

The simplest form of stand-alone PV system is a direct-coupled system, where the DC output of a PV array is directly connected to a DC load. This type of system may be combined with other generation technologies such as wind, engine-generator, or utility power as an auxiliary power source to form a PV-hybrid system. Such flexibility makes stand-alone PV systems suitable for powering houses and facilities in rural areas or remote islands. For a country like Indonesia with many people living on remote islands, stand-alone PV systems can play a role in increasing people's access to electricity.

However, most people live in areas that are connected to a public grid. Therefore, grid-connected PV systems are a more common application. A grid-connected PV system operates in parallel and interconnected with the electric utility grid. It comprises an inverter that converts the DC power from the PV array into AC power synchronized with the voltage and power quality requirements of the utility grid. Grid-connected PV systems automatically stop supplying power to the grid when the utility grid is not energized for safety reasons.

Photovoltaic system performance is defined by the performance ratio (PR). Specifically, the performance ratio is the ratio of the actual and theoretically possible energy outputs of PV systems. Before 2000 a typical PR was about 70%, while today it is in the range of 80–90% [35]. In a country like Indonesia where the grid can be weak in some areas, and which is characterized by frequent outages and fluctuating voltage, grid-connected PV systems could increase the reliability of electricity supply.

Solar irradiance is the main factor that determines the amount of electrical energy that could be potentially produced by PV systems. As such, climatic conditions determine the annual and seasonal performance of PV systems. These conditions are location dependent. Although Indonesia has a rainy tropical climate in general, the climate and weather across Indonesia are not uniform. The topography, orientation, and structure of the islands are among the influencing factors that affect Indonesia's climate [36]. Its location around the equator and oceans that surround it means that the Asian and Australian monsoons also influence Indonesia.

For climate classification, in this book, we use the Köppen-Geiger system [37]. It uses differences in temperature and rainfall as the basis of classification. Köppen-Geiger uses two or three letters to determine the climate group of a location on Earth. It divides the world into five main climate groups based on the five principle vegetation groups. The first letter represents the five climate groups, where A is for equatorial or tropical rainforest, B is for arid or dry, C is for warm temperate or moderate rainfall, D is for cold, snowy rainforest, and E is for polar [38]. Further, subgroups can be assigned as the second letter, which indicates the type of precipitation (water vapor condensation product in the atmosphere). The third letter represents the air temperature [39].

Figure 1.5 shows a map of different climates in the world according to Köppen-Geiger. It can be seen that countries around the Equator are covered by climate A. Climate B can be found in Australia, southern and northern Africa, middle southern

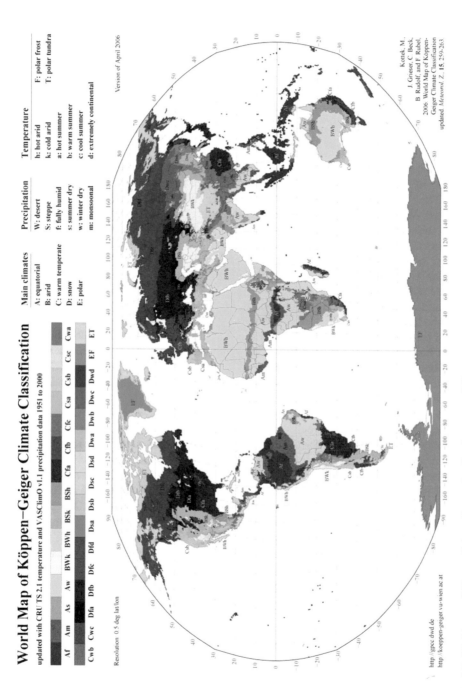

Fig. 1.5 World map of Köppen-Geiger climate classification system [51]

America, Middle-East, and western parts of the US. Climate C dominates West Europe, southern Australia, New Zealand, Japan, South-East regions of the US, Southern America, Africa, and China. Climate D covers most of Russia, Canada, and the northern US. Finally, climate E is over northern Canada, Iceland, northern Russia, and the North and South poles.

In Fig. 1.6, the Köppen-Geiger climate classification map is zoomed to Indonesia using the most recent condition available on Google Earth for 2017 [40]. It can be seen that instead of having one climate class, Indonesia has four climates. Most of the area of Indonesia is covered by red, which signifies climate A. In regions with climate A, the monthly average temperature is above 18° C, there is no winter, and substantial annual rainfall occurs exceeding the yearly evaporation [37].

Most of Indonesia's region has climate Af, which means tropical rainforest climate. The climate type Af covers almost all of the islands of Sumatra, Kalimantan, Sulawesi, Maluku Islands, and Papua. However, the climate in Java is more diverse due to two other variations: Am or tropical monsoon and Aw which means tropical savanna. The differences between Af, Am, and Aw are signified by the monthly rainfall, where Af has the highest monthly rainfall, followed by Am, and Aw is characterized by the least rain.

The tropical monsoon climate is found in parts of the west and the center of Java, the island of Bali and several regions in East Nusa Tenggara (ENT) and Sulawesi, while the tropical savanna climate can be found in the east of the island of Java and most of the regions of West Nusa Tenggara (WNT) and ENT and a little in the southeast of Sulawesi. With a higher resolution map, we could see a green line across the mountainous region of Papua, which means the oceanic sub-arctic climate (Cfc). This climate has an average temperature in the coldest month of above 0 °C and the average in other months is between 10 and 22 °C.

The average ambient temperature in Indonesia is 27 °C, with 0.03 °C annual increase [41]. From measurements made during 1981–2018, the highest ambient temperature was recorded in East Kalimantan, at 36.6 °C. Indonesia has a relatively low wind speed of around 3.8 m/s [42]. The highest average of wind speed of 16.3 m/s was observed at the province of Gorontalo in northern Sulawesi [23]. The highest rainfall was recorded in West Sumatra, with a value of 4,824.10 mm. West Java, however, is the province with the highest number of rain days, which is 295 days per year. The highest air humidity of 100% is found in several provinces, namely Jambi, Bengkulu, Riau Islands, Yogyakarta, Central and East Kalimantan, South-East Sulawesi, and Maluku. Finally, ENT is the province with the least rainfall, with the number of rain days being only 89 days throughout the year [23].

Several climatic factors, mainly solar irradiance and ambient temperature, influence energy production from a PV system. Irradiance is the power of the electromagnetic radiation on a surface and is measured in watts per square meter (W/m^2). The higher the irradiance on the module surface, the more energy is produced. The air temperature affects the efficiency of PV systems, where under a lower temperature, silicon PV modules convert solar irradiance into electricity at higher efficiency [43]. Thus, the effects of climate and system parameters on solar electricity have been proven to be significant, especially on the performance of PV plants [44–46].

Fig. 1.6 The most recent condition of Indonesia's climate available on Google Earth for 2017 [40]

Figure 1.7 shows the PV power potential in Indonesia using the variables final yield, Y_f, and reference yield, Y_r, taking into account the solar irradiance and ambient temperature [43]. The final yield is defined as the annual, monthly, or daily net alternating-current (AC) energy output in kilowatt-hour (kWh), E_{ac}, of the PV system per installed power, P_{rated}, in kilowatt-peak (kWp). The final yield can be used to compare PV plants of different systems in different climates. The reference yield (in kWh/kWp), Y_r, is the total amount of available in-plane solar irradiance in kWh/m^2, H_i, divided by the reference irradiance, $G_{i,ref}$, of 1,000 W/m^2. It can be seen that the final yield, Y_f, of silicon crystalline PV modules in Indonesia ranges from 1,095 to 1,680 kWh/kWp per year. The reference yield, Y_r, spans from 3 h/day to 4.6 h/day. The majority of Indonesia's areas have Y_r of around 3.4 h/day. However, regions of eastern Java, Bali, Nusa Tenggara, and Sulawesi have higher Y_f values. Chapter 5 will explain the potential of PV systems in Indonesia in more detail.

Besides a relatively high energy density of solar resources across the country, another good feature of the solar resource in Indonesia is that it is relatively stable across the year. This is different from locations at higher latitudes, where seasonal variation is significant.

1.6 Stakeholders' Interest in Regulations

Based on the type and structure of the government, the energy sector in Indonesia involves various stakeholders, each with different (but sometimes overlapping) roles. Figure 1.8 maps the energy stakeholders in Indonesia according to their core, direct, and indirect involvement using an example of PESTLE stakeholders' mapping on renewable energy in Indonesia [47] and the example of task distribution among the government institutions in the energy sector according to RUEN [47]. The PESTLE approach accounts for political, economic, social, technological, legal, and environmental aspects.

As shown, the central government and national level institutions play the core role in energy development in Indonesia. Their roles include enacting policy, laws, and regulations, and appointing institutions to implement the energy projects. Similarly, the other central government institutions are directly involved with financial responsibility or providing resources for the project implementation, such as land. In the implementation stage, the private electricity company and local government are usually directly involved in energy projects. Finally, indirect involvement is exercised by some other ministries, the Indonesian Chamber of Commerce and Industry, non-profit organizations and the Corruption Eradication Commission. Depending on the type of project, the public is often indirectly involved, although their direct involvement could also be seen in the community-based energy projects.

The central government makes regulations that apply nationally, while local governments can make regulations at the local level as long as they are not in contradiction with national regulations. Legal products made by the central government may include laws (UU), government regulations (PP), presidential regulations (Perpres),

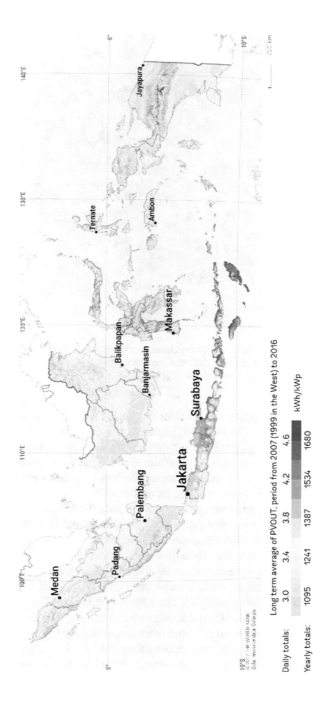

Fig. 1.7 Photovoltaic power potential in Indonesia [43]

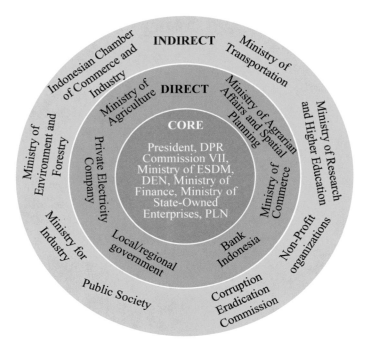

Fig. 1.8 Map of energy stakeholders in Indonesia based on the example of PESTLE stakeholders map in renewable energy in Indonesia [47] and the example of task distribution among the government institutions in the energy sector according to RUEN [47]

presidential instructions (Inpres), presidential decrees (Kepres), ministerial regulations (Permen), and ministerial decrees (Kepmen). At the provincial level, legal products can take the form of regional regulations (Perda), governor regulations (Pergub), and governor decrees (Kepgub). While at the district/city level, regional regulations (Perda), regulations of regents/mayors (Perbup/Perwako), and decrees of regents/mayors (Kepbup/Kepwako) can also be produced.

Law No. 30/2007 concerning energy serves as an umbrella for the energy sector in Indonesia [47]. It contains the main aspects of regulating and managing energy and energy sources. This law also mandates the formation of the National Energy Council (DEN), which is assigned to formulate the National Energy Policy (KEN). At present, KEN, along with other derivative regulations, will release a strategic plan for realizing the national energy resilience and independence, which is named the General Plan for National Energy (RUEN). The KEN is legalized by the government regulation number 79/2014 while RUEN is regulated in the presidential regulation number 22/2017. RUEN is the foundation and central direction for the government in deciding existing policies, including determining the roles of existing stakeholders.

According to RUEN, stakeholders in the energy sector comprise multiple parties. First, the state's ministry/non-ministerial institutions, who must take RUEN as a guide in strategic planning. Secondly, the provincial governments formalize the derivative of RUEN, namely the Regional-Provincial General Energy Plan (RUED-P). Finally, apart from government institutions, there are important stakeholders in the energy sector, such as academic and research institutions, industries, and non-profit organizations.

The National Energy Policy (KEN), which was made in 2014, has four main policies and six supporting policies. The main policies include energy availability for the national needs, priority for energy development, utilization of national energy resources, and domestic energy reserves. The six supporting policies include the following:

1. Energy conservation, energy resource conservation, and energy diversification;
2. Environment and safety;
3. Prices, subsidies, and energy incentives;
4. Infrastructure and access for the community and industry to energy;
5. Research, development, and application of energy technologies; and
6. Institutions and funding.

The Ministry of Energy and Mineral Resources (ESDM) is the leading institution that oversees the energy sector. It coordinates the current strategies, programs, and activities to produce related instruments. For example, ESDM is responsible for prioritizing energy development by putting more focus on local energy resources to increase the use of new and renewable energy (NRE). Therefore, ESDM can be very influential in renewable energy development, as discussed in Chap. 2. Other ministries are also involved in energy development, either directly or indirectly.

In addition to the above list of stakeholders, there are many other parties, such as technical consultants, service providers, associations, goods procurement services, and financial institutions.

From the research domain, academics, non-profit institutions, and industry become indispensable stakeholders in the energy sector. At present, various national and international research institutions continue to provide and publish consultancies, guidelines, and suggestions. Big names like the International Renewable Energy Agency (IRENA), the Institute for Essential Services Reform (IESR), the World Bank, and the Indonesian Institute for Energy Economics (IIEE) continue to provide their perspectives and research outcomes to improve Indonesia's performance in the energy sector. Academic publications from various universities in Indonesia and abroad are also critical in the research domain.

Some regulations regarding renewable energy tariffs and subsidies were introduced during recent years. The latest regulation is the Ministry of ESDM Regulation No. 49/2018, which allows the customers of PLN, the national utility company, to install and operate rooftop PV system for their own use and 'sell' excess energy to PLN under a net metering scheme. However, a 35% discount on the total exported energy is applied. For rooftop PV systems that belong to industrial entities, a parallel operation charge must be paid to PLN consisting of a connection charge, energy charge, and a capacity charge. The charges are based on a minimum monthly take-or-pay obligation of 40 h. The Association of Rooftop PV System Users (PPLSA) complained about this regulation which, according to them, hampers the development of rooftop PV in Indonesia [48].

1.7 Conclusions

Indonesia is a large country with significant variations in climate, population, and economy. These factors introduce challenges in power supply systems in Indonesia as well as open great opportunities for renewable energy development in general and solar energy in particular. Most people in Indonesia live on the five big islands, but a significant population can also be found on the thousands of other smaller islands. Owing to remoteness, it is often difficult and expensive to provide electricity for them by connection to the main grids. Given the size of Indonesia, which covers three time zones, the distribution of wealth and electricity infrastructure is uneven. The western and central regions are relatively more advanced than the eastern region.

A growing economy and large population can be translated into a big market for renewable energy technologies. A good solar energy resource, when combined with the right approach, could open a great opportunity for large-scale or even domestic-scale PV system businesses. The government is on the right track with a plan towards a higher fraction of renewable energy in its energy mix policy. Some regulations already exist as a basis for further improvements. However, the role of the central government in the development of the energy sector is too big, which is not favorable for building local and sectoral capacity beyond the central government.

Work remains to be done at the implementation level. Indonesia needs to include all potential and current stakeholders in a correct strategy that is beneficial to all. The central government holds important roles in initiating the changes, for example by decentralizing the energy sector to local entities.

References

1. Badan Informasi Geospasial, *Rujukan Nasional Data Kewilayahan: Luas NKRI 8,3 Juta Kilometer Persegi*, vol. 2019 (Badan Informasi Geospasial, Jakarta, 2018)
2. Indonesian Ministry of Trade, Indonesia facts and figures (2015). Available https://www.embassyofindonesia.org/index.php/basic-facts/
3. R. Hall, *Indonesia, Geology* (2009), pp. 454–460
4. N. A. Pambudi, Geothermal power generation in Indonesia, a country within the ring of fire: current status, future development and policy, Renew. Sustain. Energy Rev. (2018)
5. CIA, *The World Factbook* (2019)
6. USGS, Which country has the most earthquakes? https://www.usgs.gov/faqs/which-country-has-most-earthquakes? Accessed 17 July 2019
7. K. C. A. (Deltares) Bons et al., *Indonesia Country Water Assessment* (2016)
8. Global Forest Watch, *Forest Loss Map* (2019)
9. D.W.H. Cai, S. Adlakha, S.H. Low, P. De Martini, K.M. Chandy, Impact of residential PV adoption on retail electricity rates. Energy Policy **62**, 830–843 (2013)

10. P. Nijkamp, A. Perrels, Impacts of electricity rates on industrial location. Energy Econ. **10**(2), 107–116 (1998)
11. PwC, *Power in Indonesia* (2017)
12. PLN, *Statistik PLN 2017* (Perusahaan Listrik Negara (PLN), Jakarta, Indonesia, 2018)
13. Y. Tan, L. Meegahapola, K.M. Muttaqi, A review of technical challenges in planning and operation of remote area power supply systems. Renew. Sustain. Energy Rev. **38**, 876–889 (2014)
14. B. Mudiantoro, J. Galvez, Investing in renewable energy generation and power transmission in eastern lessons learned from ADB' s. Renew. Energy **14** (2015)
15. A. El Gammal, Photovoltaics, Tomorrow's technology available today. Clim. Action **2009**(2010), 86–89 (2010)
16. The World Bank, GDP Growth (% annual) (2017)
17. J. Hawksworth, H. Audino, R. Clarry, The long view. How will the global economic order change by 2050 ? (2017)
18. T.W. Bank, GDP per capita, PPP (current international $) (2019), https://data.worldbank.org/indicator/NY.GDP.PCAP.PP.CD. Accessed 02 Sept 2019
19. K. Brown, *Human development and environmental governance: a reality check*. Gov. Sustain. 32–52 (2009)
20. N. Singh, R. Nyuur, B. Richmond, Renewable energy development as a driver of economic growth: evidence from multivariate panel data analysis. Sustainability **11**(8), 2418 (2019)
21. The World Bank, Population, total (2019), https://data.worldbank.org/indicator/SP.POP.TOTL. Accessed 02 Sept 2019
22. The World Bank, Energy use (kg of oil equivalent per capita) (2019), https://data.worldbank.org/indicator/EG.USE.PCAP.KG.OE. Accessed 02 Sept 2019
23. BPS, *Statistical Yearbook of Indonesia 2018* (BPS-Statistics Indonesia, Jakarta, 2018)
24. Indonesia Central Bank, Inflasi (2019), https://www.bi.go.id/id/moneter/inflasi/data/Default.aspx
25. ADB, Asian Development Outlook 2019, no. April (2019)
26. D.I. Stern, The environmental Kuznets curve after 25 years (Canberra, 2015)
27. World Atlas, *The Worlds Most Densely Populated Cities* (2019)
28. S. Özokcu, Ö. Özdemir, Economic growth, energy, and environmental Kuznets curve. Renew. Sustain. Energy Rev. **72**(November 2016), 639–647 (2017)
29. IRENA and CEM, The socio-economic benefits of large-scale solar and wind energy: an econValue report, no. May (2014), p. 108
30. US PopClock, *U.S. and World Population Clock*, vol. 2019 (2019), pp. 1–5
31. G. Jones, The 2010–2035 Indonesian population projection. J. Nucl. Med. **55**(12), 9N–15N (2014)
32. ESDM, *Program Indonesia Terang Dicanangkan*, vol. 2019 (Jakarta, 2016)
33. BPS-Statistics Indonesia, *Statistik Indonesia 2018* (BPS-Statistics Indonesia, Jakarta, Indonesia, 2018)
34. BPS, *Statistik Indonesia dalam Infografis* 2018 (BPS-Statistics Indonesia, Jakarta, 2018)
35. Fraunhofer ISE, Photovoltaics Report 2019 (Freiburg, 2019)
36. S. Wirjohamidjojo, Y. Swarinoto, *Iklim Kawasan Indonesia (Dari Aspek Dinamik - Sinoptik)*
37. N. Febrianti, Perubahan Zona Iklim di Indonesia dengan Menggunakan Sistem Klasifikasi Koppen, Pros. Work. Apl. Sains Atmos. Dec 2008
38. S. Rafi'i, *Meteorologi dan Klimatologi*. Bandung: ANGKASA (1995)
39. M.C. Peel, B.L. Finlayson, T.A. McMahon, Updated world map of the Köppen-Geiger climate classification. Hydrol. Earth Syst. Sci. **11**(5), 1633–1644 (2007)
40. F. Rubel, K. Brugger, K. Haslinger, I. Auer, The climate of the European Alps: shift of very high resolution Köppen-Geiger climate zones 1800–2100. Meteorol. Zeitschrift **26**(2), 115–125 (2017)
41. BMKG, Perubahan Iklim: Tren Suhu (2019), https://www.bmkg.go.id/iklim/?p=tren-suhu. Accessed 21 July 2019
42. BPS, Kecepatan Angin dan Kelembaban di Stasiun Pengamatan BMKG, 2011–2015 (2017)

43. World Bank Group, *Solar Resource and Photovoltaic Potential of Indonesia*, no. May (2017)
44. A.L. Bonkaney, S. Madougou, R. Adamou, Impact of climatic parameters on the performance of solar photovoltaic (PV) module in Niamey. Smart Grid Renew. Energy **08**(12), 379–393 (2017)
45. P.E. Bett, H.E. Thornton, The climatological relationships between wind and solar energy supply in Britain. Renew. Energy **87**, 96–110 (2016)
46. H. Bahaidarah, S. Rehman, A. Subhan, P. Gandhidasan, H. Baig, Performance evaluation of a PV module under climatic conditions of Dhahran, Saudi Arabia. Energy Explor. Exploit. **33**(6), 909–929 (2015)
47. S.W. Yudha, B. Tjahjono, *Stakeholder Mapping and Analysis of the Renewable* (2019), pp. 1–19
48. Kontan, Permen ESDM Nomor 49/2018 Hambat Investasi PLTS Atap (2019), https://insight.kontan.co.id/news/permen-esdm-nomor-492018-hambat-investasi-plts-atap. Accessed 20 June 2019
49. D. Dalet, Indonesia. D-Maps (2007)
50. S. Kuznet, Economic growth and income inequality Simon Kuznets. Am. Econ. Rev. **45**(1), 1–28 (1955)
51. H.E. Beck, N.E. Zimmermann, T.R. McVicar, N. Vergopolan, A. Berg, E.F. Wood, Present and future Koppen-Geiger climate classification maps at 1-km resolution. Sci. Data **5**, 180214 (2018)

Chapter 2
Status and Challenges of Electricity Supply

2.1 Preface

In this chapter, we will first present the status of the general energy supply and demand in Indonesia followed by that of the electricity sector specifically. Next, we will discuss various challenges in providing proper electrical power supply to the whole population. For this purpose, we evaluate the gaps or disparities in power services from one region to another, the level of electricity consumption, and the possible impact of the current power development strategy on climate change. Then, we present power supply systems and businesses in Indonesia and show the dominant role of PLN, the national utility company. Finally, the status of renewable implementation will be discussed with particular attention to PV systems. Along with this, we argue that government regulations influence the rise and fall of renewables in Indonesia. This chapter will close with conclusions.

2.2 Status of Energy Supply and Fossil Fuel Resources

During the past ten years, the total final energy consumption (TFC) in Indonesia has increased by 20%, from 138 million tonnes of oil equivalent (Mtoe) in 2006 to 165 Mtoe in 2016. To give a context to the size of Indonesia's TFC for 2016, it was equivalent to around 14% of the TFC of the 28 European Union member countries. The calculation of Indonesia's TFC, however, does not include heat, biofuel, waste, and non-hydro renewables, while the TFC of the EU countries includes almost all energy resources.

From 2007 to 2017, the primary energy supply for Indonesia was dominated by fossil fuels [1]. The supply of oil increased from 71 Mtoe in 2007 to 82.7 Mtoe in 2017, with an annual growth rate of 1.8%. In 2007, coal contributed 37.7 Mtoe,

Table 2.1 Indonesia's fossil fuel reserves and scenarios for the time remaining for mining

Fuel	2017 reserve*	Average production growth in the past 10 years (%)	2017 production*	No. of years until completely depleted	
				Scenario A	Scenario B
Coal	24,240	8.26	461	20	52
Oil	3,170	−1.68	292	–	10
Natural gas	143	0.65	3	41	48

*Coal reserve is shown in millions of tons, oil in millions of barrels, and natural gas in a million MMSCF

which increased by 7.2 Mtoe in 2017. Similarly, the share of natural gas grew 2.4% per year during the same period.

In 2017, the proven reserves of fossil fuels were high. They covered 24.2 billion tonnes of coal, 3.2 billion barrels of oil, and 100.4 Trillion cubic feet (TSCF) of natural gas. Therefore, Indonesia strongly exploits fossil fuel and is a net exporter of coal and natural gas. Indonesia's coal production increased from 217 million tons in 2007 to 461 million tons in 2017, most of which was exported. Also, the natural gas production grew from 2.805 TSCF in 2007 to 2.963 TSCF in 2017. In contrast, oil production during the same period decreased from 348 million barrels in 2007 to 292 million barrels in 2017.

If there are no discoveries of new reserves of fossil fuels in the forthcoming period, we can estimate how long fossil fuel mining will last in Indonesia (Table 2.1). Assuming that coal mining would experience an annual production growth of 8.26% from 2018 onward, which is the average growth rate for the past ten years (Scenario A), its reserve would last for 20 years. However, if the present Reserve/Production (R/P) ratio is used (Scenario B), coal reserves would be depleted within 52 years. Using the same approach, natural gas would be completely depleted in 41 years assuming an annual production growth of 0.65% from 2018 onwards, or 48 years assuming the present R/P ratio. The oil reserve depletion rate can only be calculated using the present R/P ratio, which shows that Indonesia would already be running out of oil in 10 years' time, that is to say before 2030.

2.3 Challenges in Electricity Supply

Electricity consumption is an important factor in a country's national development and economic growth [2]. Moreover, it is an essential element for transforming economic structures and improving people's welfare [3]. Electricity consumption in Indonesia is rapidly increasing in line with economic growth [2] and population growth [4].

2.3.1 Regional Differences

At present, Indonesia is still facing challenges in providing a proper electrical power supply to the whole population. The annual growth of electricity demand in Indonesia is 6% per year. The total electricity demand in 2017 was 223 TWh, and is projected to reach 1,767 TWh by 2050, [5] or around 40.7% of the projected electricity consumption in all EU countries in 2050 [6].

Still, in 2017, 4.7% of the Indonesian population remained without electricity services [7]. The electrification ratio, ER, is the number of households with access to electricity divided by the total number of households in the study area, such as a province or a country. The number of people without electricity supply in Indonesia was equal to approximately 12.1 million people[1], which is more than the total population of Belgium [8].

The differences in ER among the provinces are also enormous. By the end of 2017, the overall ER in Indonesia was 95.4%, which ranged from the lowest value of 59.9% in ENT, the average value of around 90.7%, to the highest value of 99.9% in some provinces in Java and Sumatra (Fig. 2.1). Even, in a large city like Jakarta, some households in the islands lack access to electricity due to poverty.

The method used by the Indonesian Government for the calculation of the ER, however, can be made more accurate by redefining the term 'household with electricity.' Now, as long as a house has an electricity supply, no matter if it is only enough to turn on a small lamp, it is classified as a household with electricity. This is the case in Papua and West Papua. In these provinces, a significant number of households participated in an energy-efficient solar lamp (EESL) project in 2018. This project increased the ER of West Papua by 2.6% and the ER of Papua by 25.6%. The sustainability of the EESL project has not been evaluated yet. Once a household received an EESL, it was registered as a household with electricity, irrespective of whether the lamps remain working or not at the present time. It would be useful to redefine the criteria for households with electricity. For instance, a home is considered to have electricity if it meets an annual lower limit of electrical energy consumption, a minimum power need, or involvement in grid connectivity or diesel gensets.

A big gap in supply distribution characterizes access to electricity across Indonesia. Java and Bali are the most densely populated islands in Indonesia, as well as the centers for industry and critical economic activities. In 2017, 58.2% of Indonesia's population lived in Java and Bali, and around 68% of the national generation capacity of 54.6 GW (i.e. 37 GW) supplied Java's and Bali's demands only. On the contrary, Sumatra's generation capacity was 10.2 GW out of the total 54.6 GW, and the remaining 13.5% of the national generation capacity was located in the other two-thirds of Indonesia's area. Similarly, of the electricity produced in 2017, approximately 74% was sold in Java and Bali, 14% in Sumatra, and the rest of Indonesia, which comprises Kalimantan, Sulawesi, Papua and island provinces, received only 12% of the total national electricity production [11].

---

[1]For this estimation, we assume that the percentage of households without access to electricity corresponds linearly to the percentage of the population without electricity.

Fig. 2.1 Map of Indonesia showing ERs for 34 provinces in 2017. Data is based on Ministry of Energy and Mineral Resources (ESDM) [9], and the map is based on Dalet [10]

2.3.2 Low Electricity Consumption

Despite the size of its economy, electricity consumption in Indonesia is relatively low compared to other Asian countries and any European country. In 2017, electricity consumption per capita was only 1 megawatt-hour (MWh) [12], which was low compared to that in Vietnam at 1.6 MWh/capita, Thailand at 2.7 MWh/capita, Singapore at 8.7 MWh/capita (Fig. 2.2) [13], and Netherlands at 6.7 MWh/capita [14]. Among the reasons for the low electricity consumption per-capita in Indonesia are the big size of the population, small ER in rural areas, and low level of economic activity outside Java and Bali.

An increase in the business and industry consumers leads to a projected growth of electricity demand of around 6% per year. The emerging economies need electricity to build their industrial and business infrastructures, while the developed countries will have better opportunities to apply new energy-efficient technologies in their industries [15]. Therefore, as shown in Fig. 2.3, Indonesia and Vietnam for example, as emerging economies, have high electricity and GDP growth compared to Singapore and Australia, as developed countries. Australia even showed negative growth in electricity demand in 2015, although its GDP grew.

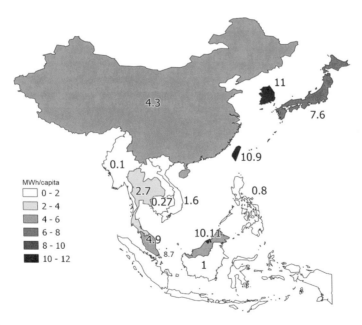

Fig. 2.2 Electricity consumption per capita in South-East Asia. The map was made using data from [12] and [13]

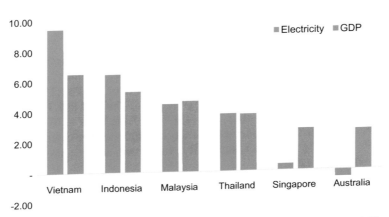

Fig. 2.3 Comparison between the increase in the average electricity consumption (left) and an increase in GDP in 2015 (right). Graphics are based on [16–18]

2.3.3 The Price Disparity Between Urban and Rural Areas

The price of electrical energy in Indonesia's main islands significantly differs from those in smaller islands. Since the end of 2017, the household electricity tariff provided by PLN, the national state utility company, is IDR 1,352/kWh ($\approx$9 US$ cent/kWh) [19]. Despite an official flat electricity price, electricity generated by diesel generators on smaller islands cost IDR 5,070/kWh ($\approx$39 US$ cent /kWh in 2015). These are typical values for diesel generators that normally run less than 12 h per day [20]. Thus, the electricity supply outside Indonesia's main islands remains expensive.

2.3.4 Imbalance in Power Sale

Most of the electricity sales take place in Java, with a figure of 2.6 times higher than for all other regions outside Java combined. By customer group, households form the largest category, using 42% of the total generated electricity, followed by industry at 32%, commercial at 19% and the public facilities at 6.7%. The total national electricity sold in 2017 was 222 TWh (Fig. 2.4) [11].

2.3.5 Variations in Reliability

Besides having access to electricity, the reliability of electricity supply is also important. Namely, an unreliable or frequently interrupted electricity supply is a major obstacle to doing business, [21] and for households, it interrupts social behavior.

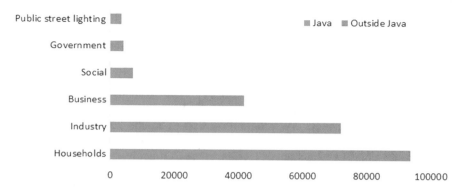

Fig. 2.4 Sales of PLN electrical energy to customer groups in Java (left) and outside Java (right) in GWh in 2017 [11]

The reliability of the power grid in Indonesia is improving. The World Bank ranked Indonesia in 33rd place (score 86.38 of 100) of all countries for ease of obtaining reliable electricity in large cities, especially Jakarta and Surabaya. Nevertheless, power supply in Indonesia is still characterized by frequent blackouts and brownouts, especially outside Java and Bali. The reliability of the power grid in Indonesia is strongly dependent on the location [22–26]. Namely, on the main islands of Java and Bali, the grid is more reliable than elsewhere, where blackouts occur daily. If Jakarta and Surabaya were taken out, the World Bank's ranking for Indonesia would drop greatly, to 86th place, below Thailand at 57, Malaysia at 36, and Singapore in 3rd place [27]. Indonesia, therefore, should be able to distribute electricity evenly and reliably, not only in big cities but throughout the country. Households respond to outages by using additional back-up systems, mostly diesel generators (genset). However, due to the required fuel supply, noise and exhaust gasses during their operation, gensets are considered to be less sustainable power sources. We will discuss the reliability issue more deeply in Chap. 3.

2.3.6 Power Infrastructure Development

In 2015, Indonesia experienced a power shortage that reached 21,000 megawatts (MW) across the country. The government of Indonesia responded to the problem with the "35,000 MW Power Plant Development Program". The program consists of the construction of 109 new power plants totaling 35,627 MW, 48,000 network kilometer (kms) transmission lines, and 114,000 megavolt ampere (MVA) sub-stations [28]. Fifty-nine projects would be constructed on Sumatra, 34 projects on Java, 49 projects on Sulawesi, 34 projects on Kalimantan, and 34 projects in Eastern Indonesia. After the successful completion of the 35 GW program, the power shortage problem in Indonesia should be completely solved. The duration of the program was set at five years starting from 2015.

2.3.7 Electrification Versus Climate Change Mitigation

Like many other countries, fossil-fueled power generation has dominated the electricity system in Indonesia. Until 2040, Indonesia is expected to remain as a coal-dependent country [29, 30]. While the GHG emissions were supposed to be reduced in Indonesia, carbon dioxide (CO_2) emissions from fuel combustion have increased by 34% from 340 Megatons (Mt) in 2006 to 455 Mt in 2016, with average annual growth of 3% [31]. The 35 GW program consists of around 20 GW [32] of coal-powered plants that are predicted to emit more than 10 Megatonne CO_2-equivalents each year or a cumulative amount of 1.4 Gigatonne CO_2-equivalents until 2035 [33].

2.4 Indonesia's Electric Power System

2.4.1 PLN's State Monopoly

Perusahaan Listrik Negara or PLN owns and operates most of Indonesia's public power infrastructure. PLN is a state-owned enterprise. As the dominant player in Indonesia's power sector, PLN occupies almost all electricity business chains from the electricity generation, transmission, distribution, to retail sales. The only link in the chain that the private sector can contribute is the generation sector. However, the share of the private sector is very small compared to PLN's generation capacity.

By the end of December 2017, the value of PLN's assets was around US$ 89 billion. The total installed capacity of PLN's power generation was 39.7 GW from 5,389 power plants. Around 72.4% of power plants were located in Java. Adding to its power plants, PLN rented a total capacity of 3 GW. The share of private electricity companies or independent power producers (IPP) was 23.75% of the total national generation capacity of 56 GW. Two of the largest IPP are subsidiaries of PLN itself, i.e. Indonesia Power with a total installed capacity of 6.47 GW and PJB with a total installed capacity of 7 GW.

PLN owns a total transmission network of nearly 49,000 kilometer-lines (kms) and distribution network (DN) of more than 1 million kms. The installed capacity of transmission substation transformers was 114,000 MVA in 1,742 substations. The installed capacity of distribution transformer substations was 60,000 MVA in 471,765 units.

In 2017, the total operating income of PLN was US$ 17 billion of which the electricity sales contribution was 96.59%, connection fees were 2.79% and other operating revenues 0.62%. PLN's coal-intensive growth plan exposes it to long-term financial risks that can be solved only by higher tariffs or long-term, and large, subsidies from the Indonesian government [34]. The government subsidized PLN by around US$ 3 billion in 2017.

The number of PLN employees by 2017 was 54,820 people. Employee productivity in 2017 was 4,070 MWh/employee and 1,242 customers/employees.

As a company that monopolizes the power sector in Indonesia amid the global change in the power sector, PLN faces some critical challenges. PLN needs to answer the following questions [34]. First, can PLN reduce its reliance on government subsidies? Second, can PLN adopt a more credible planning process? Third, how can PLN lower the risk of its Capex program and manage the major technology and market changes? Finally, does PLN recognize that long-term investors place a value on environmental performance?

2.4.2 Electric Power Infrastructure

In general, electricity system infrastructure consists of three main components, namely power generation, transmission network, and DN (Fig. 2.5).

During the past seven years, Indonesia experienced a significant improvement in power infrastructure. Within five years, about 15 GW of generating capacity, almost 16,000 kms of transmission line, 280,000 kms DN and 130,000 unit substations were added [11].

Up to 2017, the renewable energy generation capacity contributed only around 12% of the total national electricity production. Accordingly, this poses a challenge in regard to meeting the renewable energy target of 23% of primary energy from new and renewable energy sources by 2025. With only 8 years to go, the installed capacity of renewable energy should be doubled to meet the target.

Given its geography as an archipelagic country, Indonesia has 600 separate transmission systems and DN and eight large networks that, are de facto operated and owned by PLN [12]. Although IPP can construct transmission networks, usually to connect power plants in remote areas with the nearby PLN's substations, in the end, the transmission network reverts to belonging to PLN when the construction work is completed [12].

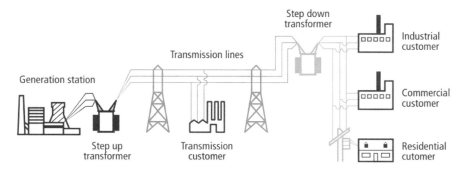

Fig. 2.5 Illustration of the electric power system [21]

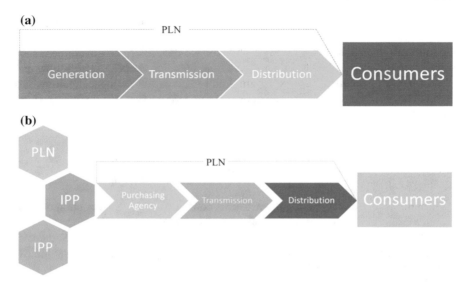

Fig. 2.6 Models in power sector business. **a** vertically integrated monopoly model, **b** Single-buyer model [35]

2.4.3 Electricity Business Models

Indonesia adopts a single-buyer model for its electricity business. Besides this model, in the global electricity industry, there are three other models, namely vertically integrated monopoly, wholesale competition, and retail competition. Before the government of Indonesia fully entrusted the power business to PLN by adopting the single-buyer model in 1985 (Fig. 2.6a), Indonesia previously used the vertically integrated monopoly model (Fig. 2.6b) [35].

2.4.4 Customer Groups

PLN divides its customers into six tariff groups, namely household, business, industrial, social, government office buildings, and public street lighting. More specifically, each tariff group is further divided based on the type of voltage, namely low, medium, and high voltage tariffs [36].

In 2017, PLN had 68 million customers, predominantly households (62.5 million). The number of commercial/business customers was 3.5 million. The number of general customers and industry were respectively 1.8 million and 76 thousand. Since 2012–2017, the total number of customers has increased by 3–4 million per year [11].

2.5 Renewable Energy

2.5.1 The Low Share of Renewable Energy

The share of renewable energy technology in the ambitious 35,000 MW program, unfortunately, is not significant. It also has not shown a growth rate as fast as it should. Renewable energy will not be able to meet the energy mix target of Indonesia, assuming a linear growth rate [37] (Fig. 2.7). In September 2018, renewable energy investment in Indonesia was only US$ 1.16 billion, which was equivalent to 57.7% of the 2018 renewable energy investment target of US$ 2.01 billion [38].

On the tariff side, the government set maximum tariff limits of 85% of local electrical energy production cost (EPC) for energy generated by PV systems. For example, if the local EPC was $ 8 cents, the energy tariff for PV systems would be $ 6.8 cents. Our interviews with PV systems project developers and investors in Indonesia indicate that such a tariff limit is a serious barrier to the development of renewable energy, which was also confirmed in [37]. In the regions with a high EPC, the renewable tariff is higher, so certain projects could be financially feasible. However, such regions usually lack infrastructure and strong networks that discourage the integration of PV systems. On the islands of Java and Bali, a tariff below 85% of BPP is considered inappropriate by the renewable energy developers [37]. Rooftop PV systems form a big market in Indonesia. However, the Decree of the Minister of Energy and Mineral Resources No. 49/2018 discourages rooftop PV systems. In this Regulation, PLN only compensates 65% of the electricity it receives from rooftop

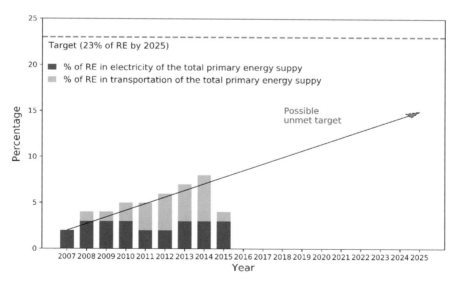

Fig. 2.7 The actual renewable energy growth in Indonesia compared to the 2025 renewable energy targets [37]

PV systems. Under this scheme, the payback period would become too long to be interesting for investors, namely 11–12 years [38].

Another burden for PV system development in Indonesia is a greater allocation of risk that must also be borne by PV system investors. The Regulation of the Minister of Energy and Mineral Resources No. 10/2018 demises the governmental force majeure, which introduces a more considerable uncertainty for the IPPs.

The last regulation that slows down the growth of renewable energy in Indonesia is the BOOT (Build-Own-Operate-Transfer) scheme. Under the BOOT scheme, the renewable developers, especially small-scale IPPs, find it is difficult to obtain funding because the BOOT prohibits developers from having collateral assets [38]. In 2017, BOOT caused 32 out of 70 signed-projects to be unable to do a financial close.

2.5.2 Challenges to PV Systems

Of all types of renewable energy, PV system utilization in Indonesia is far below its potential. Compared to the PV system potential of more than 500 GWp [38], the total cumulative installed capacity of PV systems throughout Indonesia by the end of 2017 was only about 17 MWp [38] or only 0.005% of the potential. Not only was the cumulative installed capacity low, but negative growth in the annual installed capacity was also observed since 2015, in contrast to the ongoing decline of PV system costs worldwide. As shown in Fig. 2.8, in 2015, a total of 11 MWp PV systems were installed across Indonesia. The installed capacity in the subsequent years dropped down rapidly to only 916 kWp in 2018.

The decline in the annual installed capacity of PV systems is evidence that government regulation has an enormous influence on the development of PV systems in Indonesia. We emphasize below four regulations from the Ministry of ESDM that

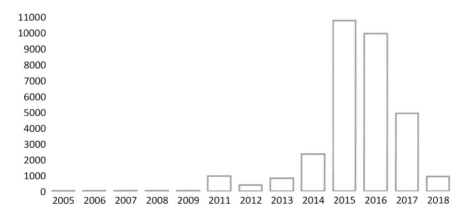

Fig. 2.8 PV system installed capacity in Indonesia, 2005–2018 (in kWp) [38]

have been slowing down the development of PV systems in Indonesia. Those regulations rule the feed-in-tariff scheme for utility PV systems that involve IPPs and the net-metering scheme for rooftop PV systems.

As shown in Fig. 2.8, the period of 2013–2016 was the golden time for PV systems in Indonesia. The reason for this boost to the PV market was due to the first Regulation of the Minister of ESDM No.17/2013 [39]. In this regulation, the feed-in tariffs for PV systems in Indonesia were introduced for the first time for IPPs. The tariff was US$ 25 cents/kWh or even US$ 30 cents/kWh if at least 40% of the systems' components were local products. As shown in Fig. 2.8, this Regulation received a positive response from the IPPs. PV systems experienced an average annual increase of 220% in the period of 2013–2015, which peaked in 2015.

However, a dramatic contrast has been seen starting in 2016 until today. Again, the main reason for this is two regulations from the same ministry, but a new minister. Namely, the Regulation of the Minister of ESDM No. 19/2016 regarding Purchasing Electric Power from PV Systems by PLN and Regulation of the Minister of ESDM No. 12/2017 regarding Utilization of Renewable Energy Resources for Electricity.

In regulation No. 19/2016, instead of using the previous single feed-in-tariff for the whole country, each PLN's regional area had its own tariff. The highest tariff of US$ 25 cents/kWh was set for the provinces of Papua and West Papua. The lowest tariff was on the island of Java, which has 5 provinces, at US$ 14.5 cents/kWh. The national median tariff was US$ 16.75 cents/kWh.

Not only did the tariff vary by regions, but Regulation No. 19/2016 also regulated the maximum capacity of PV systems that can be installed by region based on the regional electricity demand. This so-called 'capacity quota' allowed Java to get the largest quota of 150 MW because the electricity demand was higher than in other regions. The second-largest quota of 25 MW was set for the Province of North Sumatra, while the provinces of Papua and West Papua were allocated the smallest allowance of 2.5 MW due to low demand. The PV business first split into two groups in reacting to the new tariff structure under Regulation No. 19/2016; one group was negative and another group was neutral. As shown in Fig. 2.8, PV installed capacity started to decline in 2016.

Then, Regulation No. 12/2017 was enacted. In this regulation, dynamic regional tariffs were applied. Rather than using the previous fixed US$ cents/kWh per region, now, the tariff was calculated as 85% of the PLN's EPC. The PLN's EPC is determined either per PLN's regional area, PLN's distribution, PLN's system, or even PLN's subsystem. For example, in 2018, the PLN's EPC in Jakarta was US$ 6.81 cents/kWh, therefore the PV tariff is US$ 5.45 cents/kWh. The highest EPC can be found in Raja Ampat—West Papua, among others, at US$ 20 cents/kWh which gives a PV tariff of US$ 17 cents/kWh. The majority of IPPs reacted negatively to this regulation because PV systems are forced to be 15% cheaper than conventional power systems which, given the import taxes in Indonesia, will be difficult for PV systems. The consequence is clear. As shown in Fig. 2.8, the PV installed capacity, which started to decline in 2016 due to Regulation 19/2016, then dropped rapidly until 2018 due to Regulation 12/2017. An annual average decrease in the installed capacity of 140% was observed from 2016 to 2018.

Fourth, due to the sluggish market in PV business that involves IPPs, PV business in Indonesia started to shift from utility-scale to rooftop applications (net-metering) which rather than involving IPPs, works with building owners such as domestic owners and businesses. Under the previous net-metering regulation, PLN counted any single kWh exported by the PV system. The rooftop PV market grew rapidly and became the majority of the PV market in Indonesia. However, such a situation did not last long. In contrast to the Government 1 GW rooftop PV systems target in the period of 2018–2021, the PV rooftop is 'attacked' by the Regulation of the Minister of ESDM No. 49/2018. In this new regulation, of the total energy exported by a rooftop PV to the grid as recorded by the kWh meter, PLN counts only 65% rather than all energy. This changed the previous regulation that applied a 1:1 ratio between energy exported by rooftop PV systems and energy calculated by PLN.

2.5.3 Role of PV Systems

PV systems should contribute more to Indonesia's energy mix. But in fact, their share is only 0.03% of the total generated electricity. With a continuous decline in PV installation costs [38], particularly for off-grid remote applications, increasing PV capacity in Indonesia is worth considering, starting with village electricity and housing applications (Fig. 2.9).

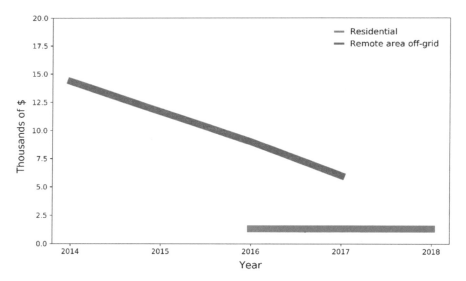

Fig. 2.9 Cost of a PV system installation (in kWp) in Indonesia, 2013–2018 [38]

2.5.4 Regulations that Support Renewable Energy

Apart from various burdensome regulations, there is also regulation that promotes renewable energy. Namely, Regulation of the Minister of ESDM Number 39 K/20/MEM/2019 [28] stimulates the acceleration of renewable energy by allowing renewable energy projects although they are not listed in the PLN's General Plan of Electricity Supply (RUPTL). Previously, no PV system could be constructed before being listed in the RUPTL. This regulation is expected to raise interest by PV system businesses to develop renewable energy projects in Indonesia [40].

2.6 Conclusions

Several challenges characterize the electricity supply in Indonesia. The regional difference between one province and another is the foremost problem. Java and Bali have the most reliable and sufficient electricity service, while Eastern Indonesia continues to have low ER and a low-reliability power supply. This also causes electrical energy price disparity between urban and rural areas. Despite its economic size and growth rate, the electricity consumption per capita in Indonesia is low compared to some other ASEAN and European countries.

Those challenges have been responded to by accelerating the development of power infrastructure across the country, such as the 35,000 MW program. Electricity consumption could be increased along with economic development and larger generation capacity. But, fossil fuel reserves are facing depletion and therefore renewable power generation is a favorable option. Although renewable energy is not a priority at the moment, it could play a crucial role in the future to combat climate change and ensure energy security.

The government and PLN, with their dominant power and authority, are expected to facilitate a fair 'playing field' for renewable power generation, which does not happen at present. Therefore, new supporting regulations are highly needed.

References

1. ESDM, *Handbook of Energy and Economic Statistics of Indonesia 2018* (Jakarta, 2018)
2. A. Sugiyono, *Pengembangan Kelistrikan Nasional Prosiding Seminar Nasional Pengembangan Energi Nuklir II* (2009)
3. L. Adam, M.T. Sambodo, Indonesia's dynamic electricity power sector: investigating need and supply performance. Econ. Financ. Indones. **61**(1), 53–68 (2015)
4. A. Rehman, Z. Deyuan, Investigating the linkage between economic growth, electricity access, energy use, and population growth in Pakistan. Appl. Sci. **8**(12), 2442 (2018)
5. BPPT, *Indonesia Energy Outlook 2018* (BPPT, Jakarta, Indonesia, 2018)
6. Eurostat, Electricity and Heat Statistics (2019), https://ec.europa.eu/eurostat/statistics-explained/index.php/Electricity_and_heat_statistics#Production_of_electricity

7. Dirjen Ketenagalistrikan ESDM, *Realisasi Rasio Elektrifikasi Akhir Tahuan 2018* (2019)
8. United Nations, World Population Prospects 2019—File POP/1-1: Total population (both sexes combined) by region, subregion and country, annually for 1950–2100 (thousands). United Nations—Department of Economic and Social Affairs—Population Division (2019)
9. Dirjen Ketenagalistrikan ESDM, *Realisasi Rasio Elektrifikasi Akhir Tahun 2017* (2018)
10. D. Dalet, Indonesia. D-Maps (2007)
11. Dirjen Ketenagalistrikan ESDM, *Statistik Ketenagalistrikan 2017* (Jakarta, Indonesia, 2018)
12. Y. Kamarudin, G. Natakusumah, S. Winzenried, T. Watson, Power in Indonesia: investment and taxation guide, no. Nov, p. 191, 2018
13. PwC, *Power in Indonesia* (2017)
14. IEA, IEA Atlas of Energy (2017), http://energyatlas.iea.org. Accessed 12 Sept 2019
15. J. Vasconcelos, Some brief remarks on security of electricity supply (2004), no. Mm, pp. 1–18
16. IEA, *Atlas of Energy*
17. OECD Development Centre, *Economic Outlook for Southeast Asia* (2015), p. 3
18. The World Bank, *GDP Growth (% annual)* (2017)
19. PLN, *Eectricity Tariff Adjustment August 2016* (PLN, Jakarta, 2016)
20. NREL, *Sustainable Energy in Remote Indonesian Grids: Accelerating Project Development* (NREL, Golden, 2015)
21. World Bank, *Getting Electricity—Factors Affecting the Reliability of Electricity Supply* (2017)
22. PLN, *Statistik PLN 2012* (PLN, Jakarta, 2013)
23. PLN, *Statistik PLN 2013*, no. 02601 (PLN, Jakarta, 2104)
24. PLN, *Statistik PLN 2014* (PLN, Jakarta, 2015)
25. PLN, *Statistik PLN 2015* (PLN, Jakarta, 2016)
26. PLN, *Statistik PLN 2016* (Perusahaan Listrik Negara (PLN), Jakarta, Indonesia, 2017)
27. World Bank, *Quality of Electricity Supply*
28. PT. Perusahaan Listrik Negara, Rencana usaha penyediaan tenaga listrik, in *Rencana Usaha Penyediaan Tenaga List* (2019), pp. 2019–2028
29. IEA, *World Energy Outlook 2015* (International Energy Agency, Paris, 2015)
30. BPPT, *Indonesia Energy Outlook 2015* (BPPT, Jakarta, 2015)
31. IEA, IEA World Energy Balances 2018 (2019), https://www.iea.org/countries/Indonesia/. Accessed 15 Jul 2019
32. ESDM, *Electricity Supply Business Plan (RUPTL) 2016–2025* (Ministry of Energy and Mineral Resources, Jakarta, 2016)
33. B. Muryanto, *Greenpeace Warns of Environmental Impact of Indonesia's 35,000 MW Project* (AsiaOne, 2015)
34. M. Brown, Perusahaan Listrik Negara (PLN): a power company out of step with global trends, Apr 2018, pp. 1–20
35. D. Setyawan, Assessing the current Indonesia's electricity market arrangements and the opportunities to reform. Int. J. Renew. Energy Dev. 3(1), 55–64 (2015)
36. PLN, *Statistik PLN 2017* (Perusahaan Listrik Negara (PLN), Jakarta, Indonesia, 2018)
37. Bridle et al., Missing the 23 per cent target: roadblocks to the development of renewable energy in Indonesia, Feb 2018
38. D. Arinaldo, J. C. Adiatma, P. Simamora, *Indonesia Clean Energy Outlook Imprint Indonesia Clean Energy Outlook* (2018)
39. Indonesia Ministry of Energy and Mineral Resources, 17 Tahun 2013, (2013)
40. Media Indonesia, Kebut Elektrifikasi dan Energi Terbarurak RUPTL PLN Diubah (2019)

Chapter 3
Experiences of End-Users of the Electricity Grid

3.1 Preface

This chapter[1] explores the reliability of electricity supply in Indonesia. On the basis of a series of user surveys at three different locations in Indonesia, namely in the islands of Sumatra, Timor, and Papua, we compare the reported indices of power reliability (SAIFI and SAIDI) and experimental results from user surveys and power measurements. These users experienced higher unavailability of power delivered by the grid than expressed by the utility-reported SAIFI and SAIDI. Therefore, for this study, new indices are introduced, namely the Perceived (P) SAIFI and P-SAIDI, which are based on the frequency and duration of blackouts experienced by the users. In this chapter, we mainly refer to statistics available for 2016.

3.2 Demand for a Reliable Electricity Supply

Access to electricity is a basic need for people. However, in some countries, not all people have access to reliable electricity from the grid. According to the International Energy Agency (IEA), around 1.2 billion people, 16% of the global population, were still without access to electricity in 2015 [1]. In Indonesia, around 35 million people remained without electricity in 2015 [2], which is more than the total population of Malaysia, its neighbor country, for the same year. Further, reliability problems, which are the focus of this study, exist in areas with access to the grid [3, 4], as also occur in Indonesia [5].

---

[1] This chapter is based on 'Kunaifi; Reinders, A. Perceived and Reported Reliability of the Electricity Supply at Three Urban Locations in Indonesia. *Energies* **2018**, *11*, 140.'

K. Kunaifi et al., *The Electricity Grid in Indonesia*,
SpringerBriefs in Applied Sciences and Technology,
https://doi.org/10.1007/978-3-030-38342-8_3

Due to various economic, technical, and political problems, a low quality of electricity supplied by the grid can be expected in some developing countries [3, 4]. This issue can be characterized by the intermittent and unreliable supply of electricity to the end-users commonly through regular grid interruptions, either planned or unplanned.

A less reliable electricity supply affects individuals and family life as well as being interrelated with the development and economic condition of a country. A low gross domestic product (GDP) results from a weak power grid in a country or vice versa. For instance, according to Murphy et al. [6], a reduction in the number of outages from 100 days per year to 10 days per year corresponds to more than a two-fold increase in GDP per person.

The reliability of electricity services can be quantified by their availability [6]. To give a preliminary impression of the reliability of the electricity supply in Indonesia as a whole, we present a calculation of the actual availability of the utility grid and the 'mean time between failure ($MTBF$)'. Availability, A, is the percentage of time that a system is functional, or the time the system is up (T_{up}), divided by the total time at risk ($T_{total} = T_{up} + T_{down}$) (Propst 1995, in [6]). The calculations follow the procedure suggested by Murphy et al. [6]. Data from the World Bank [7] are used, where the monthly average number of outages in Indonesia in 2015 is 0.5 and the mean time to repair ($MTTR$), or the average duration of a typical outage, is 5.7 h.

For the average month lasting 730 h, Indonesia's 0.5 outages per month averaging 5.7 h in duration gives a $T_{down} = (0.5 \times 5.7)$ or 2.85 h on average for the grid in Indonesia. Thus, A can be calculated, resulting in 99.6%. Also, $MTBF$ can also be calculated, since A is $MTBF$ divided by the total of $MTBF + MTTR$ [6], to be 58.1 h. These calculations show a very high reliability of electricity supply in Indonesia, while narratives of end-users indicate that the reliability might be unsatisfactory.

The reliability of electricity services is also often quantified with indices such as the System Average Interruption Duration Index (SAIDI) and System Average Interruption Frequency Index (SAIFI) [8], which are normally documented by utility companies. However, self-reported reliability indices do not always represent the actual situation accurately [6]. In this sense, knowledge about the experience of the grid users can be useful to evaluate the reality of the reliability indices of electricity service.

How reliable is the electric power supply through the distribution networks (DN) in urban areas of Indonesia from the perspective of users, and how does this compare to official data? This interesting question guides this research. Currently, information is lacking about this important topic affecting the lives of millions of people in Indonesia. To the best of our knowledge, this paper presents the first independent study for Indonesia with an evaluation of household perception regarding the reliability of the electricity supply through the distribution grid, and how the user experiences compared to the reported data from the utility.

3.2.1 Energy Demand and Electricity Supply

Given its large and growing population of 255 million people and strong economic growth, Indonesia's demand for electricity is rapidly increasing. The electricity consumption in 2015 was 201 TWh and is projected to reach 2008 TWh by 2050 [9]. However, due to the geographical distribution of the archipelago of Indonesia, the country faces challenges for providing a sufficient, evenly distributed, and reliable electrical power supply to all locations and islands (Fig. 3.1).

Java and Bali are the most densely populated islands in Indonesia as well as the center for industry and critical economic activities. In 2015, 58% of Indonesia's population lived in Java and Bali, and around 70% of the 48 GW of the national generation capacity (i.e., 34 GW) supplied Java's and Bali's demand only. The island of Sumatra used 10 GW out of 48 GW, and the remaining 9% of the national generation capacity was used in the other two-thirds of Indonesia. Similarly, of the electricity produced in 2015, approximately 75% was sold in Java and Bali, 29.3% in Sumatra, and 10% in Kalimantan and eastern Indonesia [9].

The majority of Indonesia's public power infrastructure is operated by Perusahaan Listrik Negara or PLN. PLN is a state-owned company and is the major power provider in Indonesia for electricity generation, transmission, distribution, and retail sales.

The electrification ratio (ER) is the number of households with electricity divided by the total number of households in the study area such as a province or a country. By the end of 2015, the overall electrification ratio (ER) in Indonesia was 86%, which ranged from the lowest value of 36% in Papua, a moderate value around 65% in some provinces on the islands of Sulawesi, Kalimantan, and Timor, to the highest value of 98% in Jakarta [2] (Fig. 3.1). Electrification ratios in two other provinces selected in this study, the Province of Riau and the Province of East Nusa Tenggara (ENT), were 71.5 and 52.3%, respectively. These significant disparities in the ERs of the three provinces do not represent many differences of the absolute number of households without electricity. Approximately 530,000 households without electricity could be found in ENT, 470,000 in Papua, and 435,000 in Riau.

Access to electricity in rural areas of Indonesia has increased rapidly over the past decade [11]. According to the World Bank [12], an increase of 2% per year occurred in urban areas and 20% per year in rural areas from 2004 to 2014, with respect to data from 2004. However, the IEA [13] reported an ER of 84% for 2016, which is slightly lower than the ER for 2015 based on the value from PLN of 86.2% [2]. According to the IEA, access to electricity in urban and rural areas of Indonesia in 2016 was 96 and 71%, respectively, while PLN does not differentiate between rural and urban areas in its reported ER values. Lack of access to electricity and unreliable power supply are common in rural areas, remote islands, and villages (see Fig. 3.1 and Sect. 3.2.2). However, in cities issues also exist with access and reliability of the electrical energy supply, although to a lower extent compared to elsewhere. Despite this situation, there exist few, if any, publications on the reliability of electric power or the resilience of grids in Indonesia.

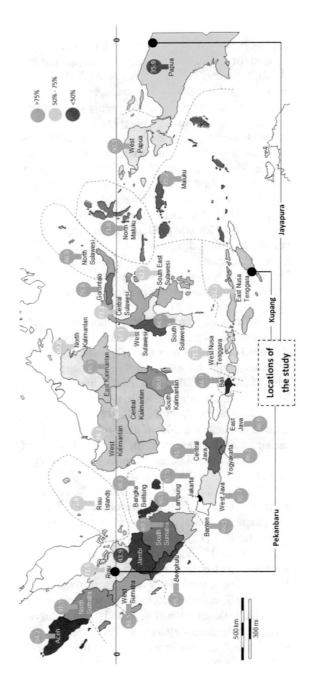

Fig. 3.1 Map of Indonesia showing ERs for 34 provinces in 2015. Data are based on Perusahaan Listrik Negara (PLN) [2], and the map is based on Dalet [10]

3.2.2 Reliability of Electric Power

Maintaining a reliable electricity supply over such an enormous distribution of islands, as the Indonesian archipelago, is a challenge due to higher investment costs for power infrastructure development. This was confirmed by Knoema [14] who ranked the power reliability of 144 countries based on electricity supply interruptions and voltage fluctuations. The report shows that the reliability of power supply in Indonesia in 2014 was 4.3 out of 7, which is slightly below the average world score of 4.5. Also, the CRO Forum [15] showed that the reliability of electric power in Indonesia was rated at 4 out of 7. Further, Erahman et al. [16] reported that Indonesia's Energy Supply Index (ESI) ranked 55th out of 71 countries during the period 2008–2013, where lower values represent a more secure energy supply. The effects of outages could be detrimental to the economy and social life. As such, studies on the quality and reliability of electricity supply are significant from a societal perspective.

Indonesia and the rest of the South-East Asian region are likely to experience an increased intensity and frequency of power disruptions in the future (Hashim [17]). Being located in the tropical zone, Indonesia is among the countries with a very high flash density and high risk of weather effects on outages (Zorro [18, 19], Bi and Qi [20], and NASA [21]). Lightning causes high impacts on Indonesia's power infrastructure and, according to Zorro [18], it is responsible for over 56% of the outages in PLN's 500 kV system, 28% in the 150 kV system, 69% in the 70 kV system, and 90% in the 20 kV system. Additional causes of power outages in Indonesia include issues for PLN in the region of 'Papua and West Papua' (PWP), which is involved in this study, related to equipment failures, vegetation, and overconsumption by the public along with other causes of outages in the local grids.

The electric power quality (PQ) is repeatedly used to specify the quality of voltage, the quality of current, the reliability of service, and the quality of power supply [22]. In this paper, we also present the actual voltage fluctuation to show the PQ of DN in the study locations. Power quality directly signifies the reliability of the electricity supply and is characterized by the probability of disturbance events [23], which, in this paper, is represented by the outage duration and frequency.

However, as mentioned above, there is a lack of information about the reliability of the electricity supply in Indonesia. Therefore, our objective is determining the actual reliability of distribution networks in urban areas of Indonesia with different ERs using the typical indicators of SAIDI and SAIFI (IEEE Standard 1366-2012).

SAIDI indicates the total duration of interruption for the average customer during a predefined period, in minutes of interruption per customer per year (Eq. 3.1) [8]:

$$\text{SAIDI} = \frac{\sum (r_i + N_i)}{N_T} \tag{3.1}$$

where r_i is the restoration time or the duration of interruption (minutes), N_i is the number of customers interrupted, and N_T is the number of customers. The subscript i represents the service area.

SAIFI indicates how often the average customer experiences a sustained interruption over a predefined period per customer per year (Eq. 3.2) [8]:

$$SAIFI = \frac{\sum(N_i)}{N_T} \tag{3.2}$$

3.2.3   User Perception to Evaluate the Reliability of Electricity Supply

Public perception has been widely used to evaluate the quality of a variety of public goods, such as agroecosystems, public policy, health, and electricity [24, 25]. Knowledge about public or end-user preference is an important input to policymaking or investment decisions as to the basis for a sustainable improvement of services provided by public utilities. Failure in defining user perception could lead to improper conclusions and inefficiencies.

Information about the user perception regarding public goods may be generated from different sources, such as expert opinion, secondary sources, the direct opinion of the users, or a combination of these. Direct information from the users is often preferable because expert opinion and secondary sources have drawbacks [26]. Namely, expert opinion may be subject to biased personal perception because they do not experience the real situation. Secondary sources are possible subject to a lack of validity when they are applied in a different context.

Information from the users may be obtained using questionnaires, face-to-face interviews, or the qualitative deliberative (focus group) method [24, 25, 27]. Also, the perception or valuation of people toward a good or service can be assessed using survey and polling methodologies [28], which provide diverse tools and approaches to performing representative public opinion measurements. In this study, we obtained information from end-users with questionnaires and face-to-face interviews to capture their experience of electricity service from the grid.

Public perception of the reliability of the electricity supply is frequently studied in many countries. Using data obtained from a survey of the users of large computers in Japan [29], customer preferences for reliable sources of electricity at the users' facilities are examined. The results of the study showed a trade-off between the reliability of the power supply and the price users paid. A more recent study from [30] applied a choice experimental method combined with a scale-adjusted latent class model to explore the valuation of electricity quality from the perspective of urban households in India. Their findings are interesting because, despite the limited preparedness of domestic users in India to pay for improved electricity quality and renewable energy, the grid users prefer state-owned distribution companies to private enterprises or cooperative societies. From another work by the same authors, which reviewed conditions in Germany [31], a different conclusion was suggested as respondents in Germany have a particularly high willingness to pay (WTP) for renewable energy.

Another study from [25] estimated households' WTP for improved electricity service in North Cyprus. They found that to avoid the cost of outages, households were willing to incur 3.6 and 13.9% increases in their monthly electricity bill for summer and winter, respectively. Other recent studies regarding user preference concerning public goods and services include [32–35]. From a review of these studies, public perception observed in a certain local setting could be different from the perception of people in other situations.

However, there are comparatively few examples of public perception used for energy research in Indonesia. Many studies which utilized public perception focused on health [36], environment [37–40], tourism [41], transportation [42, 43], and trade [44, 45]. Only two studies present public perception in the energy area in the Indonesian context, namely [46] with observations about the WTP for solar lamps and [47] which evaluated geothermal energy. However, both studies are not suitable for this work.

3.3   Research Questions and Methods

The main question of this paper is: 'How reliable is the electric power supply by distribution networks (DN) in Indonesia from the perspective of users, and how does this compare to official data?' In this context, the following sub-questions are explored:

1. What is the officially reported reliability of the power supply in Indonesia?

    This is answered through a desk study by analysis of PLN's annual reports containing SAIDI and SAIFI values for each province in Indonesia. The results of the analysis are presented in Sect. 3.4.1.

2. How do users in urban areas of Indonesia experience their power supply in practice?

    This question aims to discover the actual experiences of PLN customers in urban areas regarding the reliability of the electricity supply. To answer this question, a user study on households' experiences was executed in three locations in Indonesia, including the cities of Pekanbaru in the province of Riau, Kupang in the province of ENT, and Jayapura in the province of Papua. The results are reviewed in Sect. 3.4.2.

3. What is the actual power quality in distribution grids in Indonesia?

    This question is for providing evidence on the existence of outages and voltage fluctuation at the three study locations by conducting short-term measurements of the power quality of the PLN distribution networks. The results are outlined in Sect. 3.4.3.

    Annual reports from PLN from 2010 to 2015 [2, 48–52] were analyzed through a desk study to examine sub-question 1 of the research questions. PLN's annual reports

Table 3.1 Study locations and reasons for selection

City	Province	Regional location in Indonesia	Electrification ratios of the City's Provinces in 2015[a]	Assumption about the level of DN reliability	Period of the field survey
Pekanbaru	Riau	West	71	Best	27/03–14/04
Kupang	ENT	Central	52	Moderate	24/04–29/04
Jayapura	Papua	East	36	Worst	02/05–09/05

[a]PLN Perusahaan Listrik Negara (PLN)

contain statistics about the company's annual performance and data on distribution grid operation. These reports were published in the Indonesian language, and, therefore, it is useful to evaluate the data they contain for a broader global audience. The official data on SAIDI and SAIFI in different provinces of Indonesia were analyzed, and the trends in the reliability of the electricity supply in Indonesia were observed. The result from the desk study became the input to select three appropriate locations for this study with the lowest, medium, and highest values of reliability indices, as is explained in the next section and shown in Table 3.1.

For the exploration of sub-questions 2 and 3 of the research questions, a field research study was established from 27 March 2017 to 9 May 2017. The field research consisted of end-user studies with questionnaires and measurements of power on the grid that were subsequently performed at the three study cities (see Fig. 3.1).

The selection of the study locations was based on three criteria. First, the study locations should give a regional representation of the country. Therefore, Riau was selected among the provinces in the west, ENT in the central, and Papua in the eastern part of Indonesia. Secondly, the reliability of the power supply in the locations should range from the level of worst to moderate, to best. The ERs were used as inputs to make initial assumptions about the level of supply reliability. With an ER of 71%, it was assumed that Riau had the best reliability of power supply. Similarly, with ERs of 52 and 36%, ENT and Papua were assumed to have a moderate and the worst reliability for power supply, respectively. Table 3.1 shows the quantified data and the periods in which the field research occurred for each location.

To obtain the stated perceptions of respondents, data were collected through a structured questionnaire utilizing open-ended and closed-ended questions in combination with 'face-to-face' semi-structured interviews [53]. The stated perception data extracted from the questionnaires express the respondents' hypothetical responses about their experience regarding electricity services and willingness to pay (WTP) extra cost for improved reliability of power supply. The questionnaire contained 62 questions covering various topics regarding household experience living with and without electricity, the level of satisfaction about the electricity supply, willingness to pay a higher electricity bill, willingness to accept PV systems, and an energy use profile at home. Also, aspects such as income, gender, and profession were recorded.

For this paper, nine questions from the questionnaire were presented in Appendix A.2.

1. Would you accept an increase in your electricity bill for better electricity service?
2. How much increase in your electricity bill would you find acceptable?
3. Do you have a backup generator at home?
4. Do you experience a stable electricity voltage at home? (In this paper, the 'stable electricity voltage' is used as a general phrase in the questionnaires, which refers to a minimum level of voltage fluctuation. See also Sect. 3.4.2.)
5. Have you ever experienced a blackout at home?
6. On average, how often in a month do you experience blackouts?
7. On average, how long is the duration of the blackouts you experience?
8. At what time of day would a blackout event incur the most losses for you?
9. On average, what is the duration of a blackout that would incur economic losses for you?

According to the Theory of Value (Lancaster, in Bernués et al. [27]), the attributes or characteristics of a good or service determine its value for the individual who obtained it. To capture insight into end-user perception, respondents were asked to identify and rank each characteristic of the power supply they experienced. These rankings were translated into scores, rescaled, and averaged as presented in Sect. 3.4. The responses were used to estimate the P-SAIFI and P-SAIDI, the two new indices we defined in this paper, to estimate the reliability of the power supply based on the user experiences. P-SAIFI is the average frequency of interruption experienced by the respondents in a number of outage events per customer per year, where the initial letter stands for 'perceived'. P-SAIDI represents the user experience of the average duration of each interruption in hours per customer per year.

The P-SAIFI and the P-SAIDI are calculated by applying the mean of the frequency distribution (MFD) statistical method using the results of the user survey. The P-SAIFI is calculated based on the respondents' answers to the question: 'On average, how often do you experience blackouts in a month?' The P-SAIDI is calculated based on the users' responses to the question: 'On average, how long is the duration of the blackouts that you experience?' Standard deviations of the estimated P-SAIFI and P-SAIDI are also calculated.

Honest answers can be expected from the respondents if they believe their response could affect outcomes and if questions are associated with public goods [26], which is relevant to this study. However, users might still overstate their perception toward the questions compared to their real behavior or situation [26]. Therefore, we applied a correction factor, C, to produce more accurate values of P-SAIFI and P-SAIDI. The correction factor is based on an empirical finding by [54], who addressed the 'hypothetical bias' of people in preference-related studies using the meta-analysis statistical method. They examined data from 29 experimental studies and suggested: 'On average, subjects overstate their preferences by a factor of about 3 in hypothetical settings'. As such, the formulas applied to determine the P-SAIFI and P-SAIDI take a 30% correction factor into consideration (see Eq. 3.4).

The calculation of the P-SAIFI starts with classifying the monthly interruption frequencies experienced by the respondents, f_{Fi}, into four groups of f_{F1}: less than 3 times, f_{F2}: 3–5 times, f_{F3}: 6–10 times, and f_{F4}: more than 10 times. Then, the mid-values, x_{Fi}, of each group of interruption frequency, f_{Fi}, are found as x_{F1}: 1.5 times, x_{F2}: 4 times, and x_{F3}: 8 times. Because there is no mid-value for f_{F4}, we use an interruption frequency of 11 times to represent x_{F4}. Next, each mid-value is multiplied by the number of users, N_f_{Fi} who responded in the corresponding category, f_{Fi}, to find the frequency distribution of the data. The next step is to calculate the MFD using Eq. (3.3) [55].

$$\text{MFD}_{\text{P-SAIFI}} = \frac{\sum_{i=1}^{4} (x_{F_i} \times N\_f_{F_i})}{N}. \qquad (3.3)$$

where N is the number of respondents at each location. The values of the MFD are based on answers to the question: 'On average, how often do you experience blackouts in a month?' Therefore, they represent the number of perceived interruption events per month for each city. The final step is to multiply the MFDs by the number of months in the year. The correction factor, C, is applied to find the average annual P-SAIFI.

Equation (3.4) is built from the above steps and is used for calculating the P-SAIFI:

$$\text{P-SAIFI} = C \times \frac{\left\{ \sum_{i=1}^{4} (x_{Fi} \times N\_f_{Fi}) \right\} \times 12}{N}. \qquad (3.4)$$

where the constant, C, is the correction factor and i represents the outage frequency groups.

Similarly, calculation of the average annual P-SAIDI for each customer is based on answers to the question: 'On average, how long is the duration of the blackouts that you experience?' It starts with categorizing the monthly outage duration experienced by the respondents, f_{Di}, into five groups of f_{D1}: less than 5 min, f_{D2}: 5–15 min, f_{D3}: 15–60 min, f_{D4}: 1–2 h, and f_{D5}: longer than 2 h. Then, the mid-values of each group of interruption durations, x_{Di}, are calculated as x_{D1}: 2.5 min, x_{D2}: 10 min, x_{D3}: 37.5 min, and x_{D4}: 90 min. Because there is no mid-value for f_{D5}, an outage duration of 125 min is used to represent x_{D5}. Next, each mid-value is multiplied by the number of users, N_f_{Di}, who responded in the corresponding category, f_{Di}, to find the frequency distribution of the data. The next step is to calculate the MFD using Eq. (3.5):

$$\text{MFD}_{\text{P-SAIDI}} = \frac{\sum_{i=1}^{5} (x_{Di} \times N\_f_{Di})}{N} \qquad (3.5)$$

The final step is to calculate P-SAIDIs by multiplying the MFDs by the above P-SAIFI, and the results are divided by 60 to obtain the number of hours of interruption per customer.

Equation (3.6) is used for calculating the P-SAIDI:

$$P\text{--SAIDI} = \frac{\sum_{i=1}^{5}(x_{Di} \times N\_f_{Di}) \times P\text{--SAIFI}}{N \times 60} \tag{3.6}$$

Also, the standard deviations, s, are presented using Eq. (3.7) [56]:

$$s = \sqrt{\frac{(\sum_{i=1}^{5} N\_f_i \times x^2) - (\sum_{i=1}^{5} N\_f_i \times x)^2}{N}} \tag{3.7}$$

For power measurements, a 3169-21 Clamp-On Power HiTester (Hioki, Nagano, Japan) was installed on three-phase main distribution panels at office buildings in urban areas of the three locations. The Hioki device measured many power quality parameters, including the voltage level, with a recording interval of 1 min. The accuracy of the voltage level measurement is $\pm 2\%$.

In Pekanbaru, measurements were performed for 15 days at the office building of the Faculty of Science and Technology of UIN Suska Riau University. In Kupang and Jayapura, the local Bureau of Meteorology offices hosted the measurements for five and seven days, respectively. Measurements in Kupang covered workdays only, while in Pekanbaru and Jayapura both weekdays and weekends were included. During measurements, each office ran with their usual routine from 7:30 a.m. to 4:30 p.m. After working hours, only some lamps and measurement equipment were in operation. Figure 3.2 shows the measuring instruments used and the connection points.

Fig. 3.2 Hioki 3169-21 Clamp-On Power HiTester measuring power quality (PQ)

3.4 Results

3.4.1 SAIDI and SAIFI

Table 3.2 and Fig. 3.3 show the SAIDI and SAIFI at various locations in Indonesia based on averaged historical data from 2010 to 2015 in the Statistic PLN [2, 48–52]. Regarding SAIDI, as shown in Table 3.2, each customer in Riau experienced 11.8 h of outage per year or around 59 min per month. In Papua and ENT, each customer experienced shorter outage durations of 38 and 17 min/month, respectively, during the same period. Regarding the SAIFI, customers in Riau and Papua experienced outages more often (7.9 events per year) compared to those in ENT (6.1 events per year).

Using SAIDI and SAIFI values in Table 3.2, it can be concluded that among the three provinces, Riau has the worst reliability of electricity supply, followed by Papua in the middle with customers in ENT having the highest reliability level. This contradicts the initial assumption by the authors, as shown in Table 3.1 that Riau would have the best reliability of power supply among the three provinces. Also, it is somewhat surprising that the reported SAIDI for Papua in 2015 was only 1.4 h per year, which contradicts the narratives of end-users. Therefore, we conclude that these official figures could be questionable.

For further illustration, in Fig. 3.3, the SAIDI and SAIFI are shown in small graphs for the eight regions of Sumatra, WNT, Kalimantan, Sulawesi and Maluku, Java and Bali, ENT, Papua, and the 'Riau and Riau Islands'. As shown, from 2010 to 2015, the SAIDI slightly decreased in Sumatra and Kalimantan, whereas in Papua, they decreased rapidly. Conversely, the SAIDI in Java and Bali as well as for the whole of Indonesia only changed slightly during this period, whereas in Sulawesi and Maluku a significant increase of SAIDI took place in 2015. It can be concluded that outside the islands of Java and Bali, the SAIDI and SAIFI can vary strongly depending on the location and the year of reporting because in Indonesia the majority of the power production capacity is located on Java and Bali.

3.4.2 Household Experiences

For the user survey, 300 questionnaires were distributed in the cities of Pekanbaru (Riau Province), Kupang (ENT Province), and Jayapura (Papua Province) with an average response rate of 68%. The target households were selected randomly, but the respondents were required to be household members who are responsible for the electricity service at home, such as contracting and payment. In Pekanbaru, 114 questionnaires were filled out in 19 days, in Kupang 65 questionnaires in 6 days, and in Jayapura 26 questionnaires in 8 days.

The statistics of the respondents could be improved by increasing the quantity and having a more equal distribution over the three cities and other demographic variables

Table 3.2 SAIDI and SAIFI in Riau, ENT, and Papua, 2010–2015

Regional Area	2010		2011		2012		2013		2014		2015		Average 2010–2015	
	SAIDI	SAIFI	SAIDI	SAIFI	SAIDI	SAIFI	SAIDI	SAIFI	SAIDI	SAIFI	SAIDI	SAIFI	SAIDI	SAIFI
Riau and Riau Islands	22.9	9.0	11.6	6.5	3.9	3.2	7.2	6.7	14.1	12.5	11.1	9.6	11.8	7.9
Papua	16.8	16.1	15.2	14.5	7.6	9.5	3.0	3.1	1.7	2.3	1.4	1.8	7.6	7.9
ENT	4.2	7.8	3.1	5.3	4.5	7.5	3.3	5.5	3.9	4.9	4.5	5.6	3.9	6.1
Indonesia	7.0	6.8	4.7	4.9	3.85	4.22	5.76	7.26	5.81	5.58	5.31	5.97	5.4	5.8

SAIDI is given in average hours of outage duration per customer per year, SAIFI is given in average outage events per customer per year. Data sources: PLN (2011–2016) [2, 48–52]

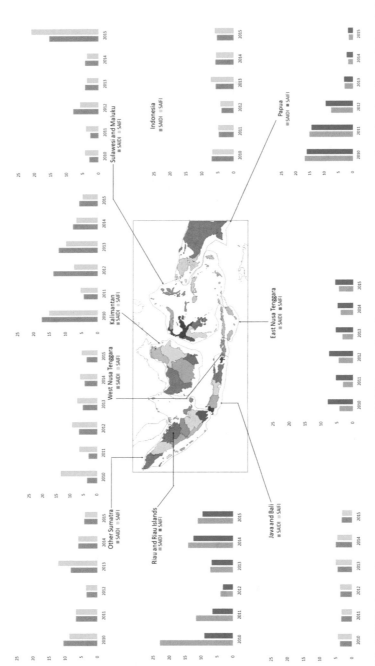

Fig. 3.3 SAIDI in hours of outage per customer per year and SAIFI in a number of outage events per customer per year for various locations in Indonesia for 2010–2015. Data sources: PLN (2011–2016) [2, 48–52]

2.3.2   Low Electricity Consumption

Despite the size of its economy, electricity consumption in Indonesia is relatively low compared to other Asian countries and any European country. In 2017, electricity consumption per capita was only 1 megawatt-hour (MWh) [12], which was low compared to that in Vietnam at 1.6 MWh/capita, Thailand at 2.7 MWh/capita, Singapore at 8.7 MWh/capita (Fig. 2.2) [13], and Netherlands at 6.7 MWh/capita [14]. Among the reasons for the low electricity consumption per-capita in Indonesia are the big size of the population, small ER in rural areas, and low level of economic activity outside Java and Bali.

An increase in the business and industry consumers leads to a projected growth of electricity demand of around 6% per year. The emerging economies need electricity to build their industrial and business infrastructures, while the developed countries will have better opportunities to apply new energy-efficient technologies in their industries [15]. Therefore, as shown in Fig. 2.3, Indonesia and Vietnam for example, as emerging economies, have high electricity and GDP growth compared to Singapore and Australia, as developed countries. Australia even showed negative growth in electricity demand in 2015, although its GDP grew.

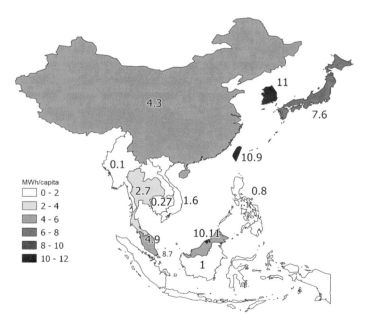

Fig. 2.2 Electricity consumption per capita in South-East Asia. The map was made using data from [12] and [13]

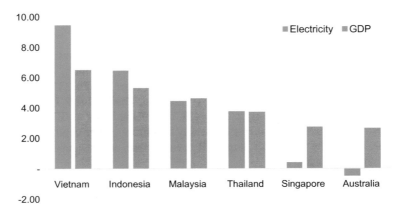

Fig. 2.3 Comparison between the increase in the average electricity consumption (left) and an increase in GDP in 2015 (right). Graphics are based on [16–18]

2.3.3   The Price Disparity Between Urban and Rural Areas

The price of electrical energy in Indonesia's main islands significantly differs from those in smaller islands. Since the end of 2017, the household electricity tariff provided by PLN, the national state utility company, is IDR 1,352/kWh (≈9 US$ cent/kWh) [19]. Despite an official flat electricity price, electricity generated by diesel generators on smaller islands cost IDR 5,070/kWh (≈39 US$ cent /kWh in 2015). These are typical values for diesel generators that normally run less than 12 h per day [20]. Thus, the electricity supply outside Indonesia's main islands remains expensive.

2.3.4   Imbalance in Power Sale

Most of the electricity sales take place in Java, with a figure of 2.6 times higher than for all other regions outside Java combined. By customer group, households form the largest category, using 42% of the total generated electricity, followed by industry at 32%, commercial at 19% and the public facilities at 6.7%. The total national electricity sold in 2017 was 222 TWh (Fig. 2.4) [11].

2.3.5   Variations in Reliability

Besides having access to electricity, the reliability of electricity supply is also important. Namely, an unreliable or frequently interrupted electricity supply is a major obstacle to doing business, [21] and for households, it interrupts social behavior.

to minimize bias. However, at the remote location of Jayapura, it was challenging to involve end-users due to transportation constraints and low population density. Since the number of respondents in Jayapura is significantly lower than those in Pekanbaru and Kupang, information from Jayapura appears to be less significant in this study, although it remains valuable as complementary information. Thus, the results of the user study for Jayapura are presented differently and shown in italic fonts in the tables and with slightly transparent color in the figures.

The demographics of the respondents are outlined in Appendix A.1 (Table A.1), and the distribution of respondents by city address is shown in Fig. 3.4. Most of the respondents in Pekanbaru and Jayapura were upper-middle-income households, but in Kupang, they originated from lower-middle-income groups. A significant number of high-income households also participated in Pekanbaru (respondents are classified into four groups of income based on the World Bank criteria (2016); for 2016, low-income economies are defined as those with gross national income (GNI) per capita of $85 or less in 2015, lower-middle-income between $86 and $335, upper-middle-income between $336 and $1040, and high-income economies are those with a GNI per capita of >$1040 or more).

Also, in the three cities, most respondents (54–63%) were aged 30–49 years followed by the age group of 50–64 years. In Jayapura, a significant number of younger respondents with an age of 18–29 years participated. Regarding the level of education, most of the respondents were well educated, which means they attended high school or higher education. However, in Jayapura, 85% of respondents were postgraduate degree holders because the questionnaires were delivered at a university. Finally, the respondents were classified as citizens living in urban-core or sub-urban areas. Urban cores are the most densely populated areas in a city with an average population density of 1000 persons/km^2, while sub-urban areas are those with 60% lower population density or less. In Kupang and Jayapura, most respondents live in sub-urban areas (55% in Kupang and 77% in Jayapura), whereas in Pekanbaru, 47% of participants live in sub-urban areas.

The first set of questions on the survey considered the importance of reliable electricity supply to the respondents. They expressed the importance of reliable electricity supply by their willingness to pay a higher electricity bill to obtain better electricity service and by ownership of a backup generator at home.

It was shown by the results of the survey that most of the respondents recognize that electricity is an important good for them. Because of the continuity of delivery of electricity is vital to them, respondents are willing to pay more for better electricity services or to buy and operate gensets. This is represented by more than half of respondents being willing to pay a higher electricity bill for better electricity services with 68% in Kupang and 56% in Pekanbaru (Fig. 3.5a and Appendix A.3 (Table A.2)).

To avoid the cost of outages, households in Pekanbaru and Kupang are willing to bear a 10–30% increase in their monthly electricity bill (Fig. 3.5b). Based on the data from PLN PWP, the average electricity expenditure of households in the urban-core of Jayapura is IDR 350,000 ($\approx$$27) per month. Using this value and assuming a similar monthly cost of electricity for households in Pekanbaru and Kupang, it can be estimated that households are willing to pay $3–$8 extra (above

Fig. 3.4   The distribution of respondents by city address in the study locations

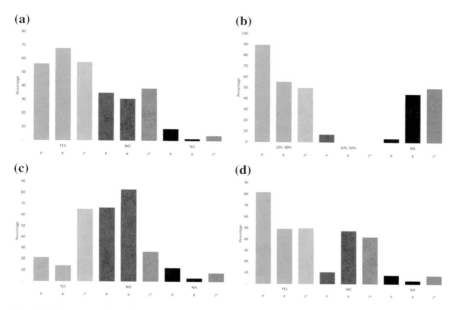

Fig. 3.5 The proportion of respondents' answers to survey questions in percentages: **a** Would you accept an increase in your electricity bill for better electricity service? **b** How much increase in your electricity bill would you find acceptable? **c** Do you have a backup generator at home? **d** Do you experience a stable electricity voltage at home? *P* Pekanbaru, *K* Kupang, *J* Jayapura *, *NA* No answer. * Data from Jayapura is less representative

their average monthly expenditure) per month or $1c–$3c per kWh for improved reliability of their electricity supply. Regarding the possession of a backup generator at home, the majority of respondents in Pekanbaru and Kupang do not have gensets, but in Jayapura, 65% of respondents do (Fig. 3.5c). However, 21% of respondents in Pekanbaru have gensets with only 14% in Kupang.

Regarding the voltage stability experienced, the survey results show that 82% of households in Pekanbaru experience stable electrical voltage in their homes compared to 49% in Kupang (Fig. 3.5d). This is based on visual observations by the users in the form of a decrease in the brightness of lamps or sudden changes of the television screen's light output. Indeed, changes in appliance behaviors could also be caused by problems with the appliances themselves or due to human errors. Therefore, visual observation takes only temporary and repeated changes in appliance behaviors into account.

The final set of questions focuses on the respondents' experiences with blackout events at home. As shown in Fig. 3.6a, most respondents in Pekanbaru and Kupang experience three to five blackouts or less per month. However, in Kupang, 31% of the respondents experience six to ten blackouts per month.

Only two households of all the respondents in the three cities stated that they never encountered any outage. The average duration of each blackout event typically is 1–2

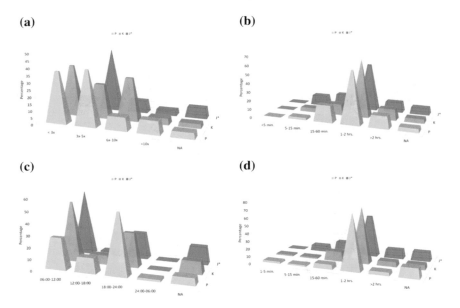

Fig. 3.6 The proportion of respondents' answers to the following questions, in percentages: **a** On average, how often do you experience blackouts in a month? **b** On average, how long is the duration of the blackouts that you experience? **c** At what time of day would a blackout event incur the most losses for you? **d** On average, how long is the duration of a blackout that would incur any losses for you? *P* Pekanbaru, *K* Kupang, *J* Jayapura *, *NA* No answer. * Data from Jayapura is less representative

h for more than half of the respondents in each city (Fig. 3.6b). Around 12–15% of respondents experienced an outage with a duration of more than 2 h each. None of the respondents experienced a blackout for less than 5 min.

If a blackout reaches 1 h in length, it begins to incur losses for most of the respondents (Fig. 3.6d). The timings of interruptions that could incur losses for users are those occurring between 6:00 a.m. and 12:00 p.m., and 6:00 p.m. and 12:00 a.m. Within these periods, electricity is highly required for work, business, and domestic activities.

To compare the officially reported SAIDI and SAIFI as presented in Sect. 3.4.1 with the user experiences, two new indices of the reliability of power supply, P-SAIDI and P-SAIFI, are defined as described in Sect. 3.3. The calculations of the P-SAIDI and the P-SAIFI are based on the results of the user survey (Appendix A.3 (Table A.2)).

Figure 3.7 and Table 3.3 summarize the results of the P-SAIDI and P-SAIFI calculations. In Table 3.3, standard deviations are provided to the right of the estimated P-SAIDI and P-SAIFI. P-SAIDI in Pekanbaru and Kupang is 21 and 24 h/customer per year, respectively. Compared to SAIDI values from PLN for the respective provinces, the P-SAIDI in Pekanbaru and Kupang are 2.6-fold and 3.9-fold higher

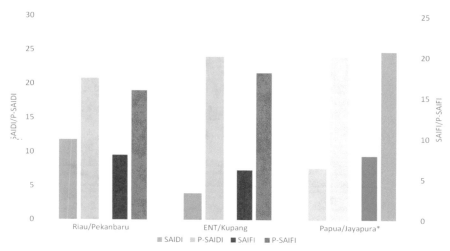

Fig. 3.7 SAIDI versus P-SAIDI and SAIFI versus P-SAIFI for the three locations

Table 3.3 Perceived and reported SAIDI and SAIFI

Parameter	Unit	P	K	J[a]
SAIDI	Hours/customer per year	7.9	6.1	7.9
SAIFI	Number of interruption/customer per year	11.8	3.9	7.6
P-SAIDI	Hours/customer per year	21 ± 33	24 ± 33	24 ± 39
P-SAIFI	Number of interruption/customer per year	16 ± 3	18 ± 3	21 ± 3

[a]Data from Jayapura is less representative

than PLN's SAIDIs. Also, P-SAIFI in Pekanbaru and Kupang are 16 and 18 outage events/customer per year, respectively. This corresponds to 1.3 times and 4.6 times higher than the PLN SAIFI for the respective provinces. Because statistical extrapolation was used to find the P-SAIDI and the P-SAIFI, the s of P-SAIDI is 33 h/customer per year in Pekanbaru and 33 h/customer per year in Kupang. The s for P-SAIFI in Pekanbaru is three interruptions/customer per year and 3.1 interruptions/customer per year in Kupang.

3.4.3 The Reliability of the Electricity Supply According to Measurements

As described in Sect. 3.3, the voltage at the distribution grid was recorded at 1 min intervals for 15 days in Pekanbaru, 5 days in Kupang, and 7 days in Jayapura. Figure 3.8 shows the time series of voltage measurements at the three locations.

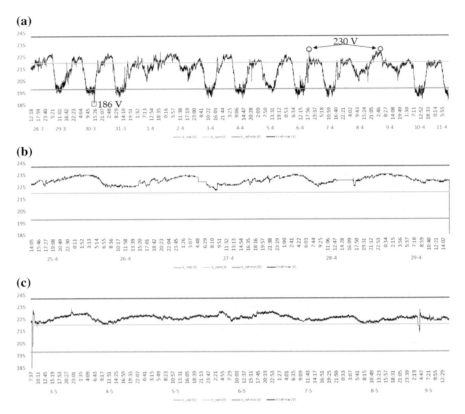

Fig. 3.8 The voltage level on the grid during the measurement period in **a** Pekanbaru, **b** Kupang, and **c** Jayapura. Note the different time scales between the measurement locations. Red line: Highest limit for voltage, Yellow line: Average voltage, and Green line: Lowest limit for voltages

As shown, based on the Ministry of Energy and Mineral Resources guideline [57], the average nominal voltage is 220 V, the highest allowable voltage limit is 242 V, and the lowest allowable limit is 198 V. The voltage level at the measurement point in Pekanbaru tends to be lower than the average nominal voltage (Fig. 3.8a). During 15 days of measurement in Pekanbaru, there were 10 days when the voltage dropped below the lowest allowable voltage limit. Unlike in Pekanbaru, during the 5-day and 7-day measurement in Kupang (Fig. 3.8b) and Jayapura (Fig. 3.8c), voltages were always within the allowable limits, although they tended to be higher than the nominal voltage, except for 2 days where the voltage in Jayapura was lower than the average nominal voltage. Figure 3.9 presents blackout events in the three locations at which power was measured at the distribution grid. Several blackout events with different durations occurred in each city during the relatively short measurement period.

(a)

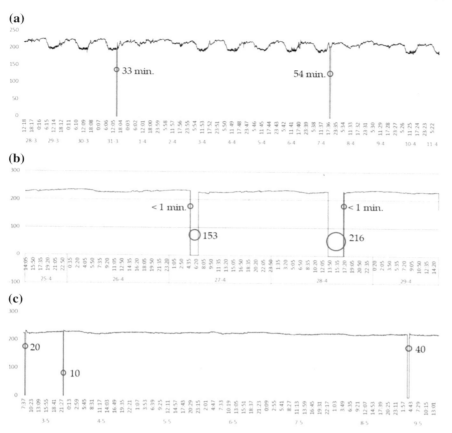

Fig. 3.9 Blackout events during the measurement period in **a** Pekanbaru, **b** Kupang, and **c** Jayapura. Note the different time scales between the measurement locations. Red circles indicate blackout events and their duration in minutes

It is shown in Fig. 3.9a that two blackouts were recorded in Pekanbaru during the 15 days of measurement. Outage events in Pekanbaru lasted 33 min or longer. Further, in Kupang (Fig. 3.9b), within a shorter period of measurement of 5 days, two blackout events were observed. Outage events in Kupang lasted longer than those in Pekanbaru (more than 153 min). Also, in Jayapura, three blackouts were captured within 7 days of measurement, ranging from 10 to 40 min. By combining the daily-averaged outage durations from measurements in the three cities, it can be concluded that the outage duration in Kupang was worst, with 74.2 min of outage per day on average, whereas in Jayapura only 10 min of outage per day was measured and only 5.8 min per day in Pekanbaru. It must be said here that these outage duration values are very high compared to many other places in the world and Indonesia. Also, it is useful that the findings from Fig. 3.9 were confirmed by the results of the user survey (see Sect. 3.4.2).

3.5   Discussion and Conclusions

This paper explores end-user experiences regarding the reliability of electricity supply in their homes and compares the reliability indices reported by the national
utility company for the cities of Pekanbaru in the province of Riau, Kupang in the
province of ENT, and Jayapura in the province of Papua, Indonesia. The research
was conducted using a desk study and a user survey.

The results of the desk study can be seen in Sect. 3.4.1, which reviews SAIDI,
SAIFI, and ER. Using a regression analysis, it was shown that there is a significant
positive relationship between SAIDI and SAIFI, $r(5) = 0.85$, $p < 0.0015$ (Fig. 3.10)
(the 'r' is the correlation coefficient of two variables for which values range from -1.0
to $+1.0$; the closer r is to $+1.0$ or -1.0, the closer the two variables are related; there
is no evidence of correlation if r is close to 0. A positive linear correlation exists if r is
positive, and a negative linear correlation exists if r is negative. The 'p' value is used to
check whether the calculated 'r' is significant. If $p < 0.05$, then the result is statistically
significant, and if $p > 0.05$, then the result is non-significant). However, SAIDI and ER
and SAIFI and ER show less significant negative correlations, respectively, with $r(5)$
$= -0.1$, $p < 0.8$ and $r(5) = -0.3$, $p < 0.5$. Therefore, the ER values of a province
cannot be used as an indicator of the level of reliability of the power supply.

In Sect. 3.4.2, the results from the user study were presented with a comparison
of the reported and perceived SAIDI and SAIFI. Also presented are the WTP, households' incomes, and genset possession. The results from the user study show there are
significant gaps between the official and perceived reliability indices. The implication
of these gaps is clear: the reported reliability indices do not always demonstrate the
experience of the grid users. It is obvious that both reliability indices, those reported
by PLN and those introduced in this study, have advantages and drawbacks.

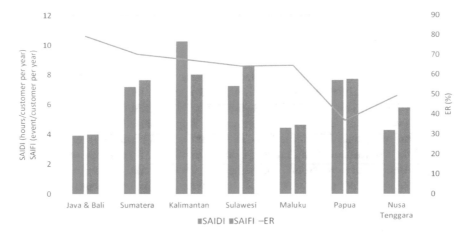

Fig. 3.10   Correlations between SAIDI (blue bar) and SAIFI (orange bar), the electrification ratio
(gray line), and SAIFI and electrification ratio

On the one hand, the PLN SAIFI and SAIDI data are likely generated based on careful documentation of actual outages, which could offer more reliable information. However, because they are based on large provincial areas ranging from 664 to 154,000 km^2, they do not distinguish between urban and rural areas. On the other hand, the perceived reliability indices introduced in this study are based at the city level with smaller resolution areas of 180–936 km^2, which could result in better accuracy. However, our indices are based on the user's perspective, which could be biased, but it can be enhanced using a correction factor as presented above. Further, it is interesting to observe the relations between different variables, such as P-SAIDI, P-SAIFI, households' income, and WTP (Fig. 3.11). Relations are shown in the scatter graphs (Fig. 3.11), which were generated using the 'randbetween' function in Microsoft Excel based on ranges of corresponding data originated from the questionnaires (because the monthly incomes of some respondents can reach above $1500, whereas the values of other variables are much lower, the household incomes in Fig. 3.11 are shown in multiples of 10 to allow for the same scaling on other variables.). This approach was also used by [58–60]. As shown, there are strong positive correlations between P-SAIDI and P-SAIFI in Pekanbaru and Kupang with the R^2 values of 0.8 and 0.6, respectively, as is also valid for the correlation between the reported SAIDI and SAIFI (Fig. 3.10).

As shown in Fig. 3.11, P-SAIDI has weak negative correlations to household income in Pekanbaru and Kupang, with R^2 values of −0.06 and −0.01, respectively. Our analysis found that high-middle income and high-income households experience a slightly higher P-SAIDI than those from low-middle and low-income households. In Kupang, higher P-SAIDI are experienced by low-income households, which are followed by high-middle income households, but low-middle income and high-income households experience fewer outages than the other income groups.

A weak positive correlation can be found between P-SAIDI and WTP in Pekanbaru with the R^2 value of 0.14. However, in Kupang, a weaker correlation between the two variables exists with the R^2 value of −0.002. This finding is somewhat similar to the results of a study by Sagebiel and Rommel (2014) in India [30]. An interesting finding of our analysis shows that in Pekanbaru and Kupang more low-medium income households are willing to pay extra for improved power reliability, although this income group experiences fewer outages. The WTP among high-income households is rather low. Even in Kupang, low-income households have higher WTP than high-income households. This is because high-income households often own gensets, which incur an extra cost of operation and maintenance, and this makes them rather reluctant to spend even more for improved electricity service. The increase in the monthly electricity bill for improved electricity service in the cities in Indonesia of 10–30% is somewhat higher than those in other countries. However, the values in dollars, a $3–$8 increase for a comparable outage duration, can also be found in other countries, such as Cyprus [25] and Sweden [61]. Therefore, it can be concluded that the WTP is a factor of the outage duration [61] as well as the factor of genset ownership. The latter is due to the assumption among respondents that 'improved reliability' does not imply a complete escape from outages, which means an expenditure for gensets and other costs may still be needed. This conclusion agrees with

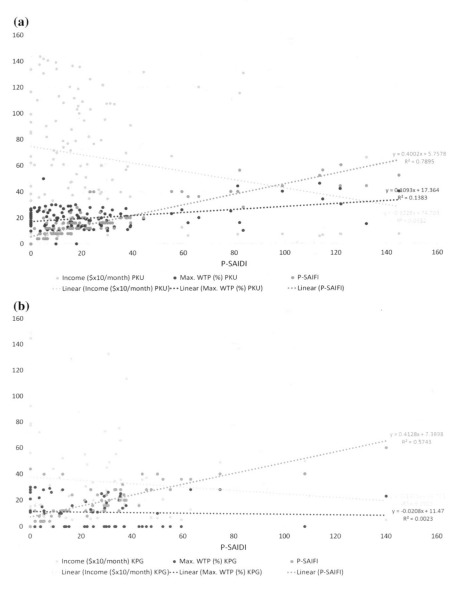

Fig. 3.11 Correlations between P-SAIDI and P-SAIFI, the income of households, and WTP in **a** Pekanbaru and **b** Kupang

[6] that users demanding more reliable electricity can expect an increased, and 'justifiable', cost of energy. This is because to achieve the desired level of reliability, additional costs to add other equipment to the existing systems is needed. However, financial incentives, which are usually provided for renewable electricity, could help customers deal with additional costs to achieve a reliable electricity supply [62].

The use of P-SAIDI and P-SAIFI indicators is relevant because it evaluates smaller areas of cities with known distribution network configurations. Measurements in the distribution grids verified the results from the user surveys. Because of the use of radial configurations in the local distribution networks, low reliability can be expected. The radial configuration offers a simple topology and is cost-effective, but has lower reliability [63]. The local nature of this study, therefore, could represent the situation in a larger area because depending on the point where an interruption occurs in the radial network, it could affect a larger area along the network lines. However, a more detailed study concerning factors that contribute to the low reliability of the distribution networks in the Indonesia supply would be valuable.

The method applied in calculating SAIDI and SAIFI and the approach to estimate P-SAIDI and P-SAIFI are relatively similar. SAIDI and SAIFI are calculated by dividing the total minutes of outages and the total frequency of outages, respectively, in the study area by the number of customers served by the grid (see Eqs. 3.1 and 3.2). The final results of the calculations are outage duration per customer per year, for SAIDI, and outage frequency per customer per year, for SAIFI. Therefore, it is not surprising that the unit of SAIDI is the same as the unit of P-SAIDI, as also applies to the units of SAIFI and P-SAIFI. The main difference between the two types of indices is the source of data input for the calculations. The SAIDI and SAIFI calculations use data collected by the utility, whereas the calculations of P-SAIDI and P-SAIFI are based on the end-user experiences.

Like all studies involving field research, our research also includes limitations due to the availability of time and person-hours for the execution of the research at three locations geographically located far from each other. This situation and a few other factors that could affect our results are explained as follows:

1. While the values for SAIDI and SAIFI are provincial values, our study occurred at specific locations, namely in the capital cities of the three provinces. However, it is still surprising that the officially reported values of reliability indices are lower than those recorded by measurements on the grid and reported by end-users. In fact, higher values can be expected for the officially reported SAIDI and SAIFI because they cover the whole province, which contains urban areas, with a relatively high-quality supply of electricity, and rural areas, which tend to have a lower power quality. It would, therefore, be valuable to expand this study to other locations in these provinces and other provinces to produce more evidence. This study offers a good start, which might be continued by other researchers.

2. The 26 respondents in Jayapura are not a representative sample for this study. Therefore, given the importance of the opinions from the end-users regarding the reliability of the electricity supply, the number of respondents in a future follow-up study should be increased to improve the statistics. Then, the statistic of

respondents could also be more equally distributed over the three locations, levels of income, and other demographic variables to minimize any bias. Similarly, the duration of the measurements of the electricity grid could be increased.

3. In our study, an income-bias of the respondents was not accounted for. As a result, this study mostly represents high and upper-middle-income classes.

However, apart from these potential points of improvement, this study is unique and fills a void in existing real-life data on experienced power quality, and it confirms our initial hypothesis that the reported indices of the reliability of the power supply from PLN are lower than the user experience. Using the Perceived-(P) SAIDI and P-SAIFI we introduced for the provinces of Riau, ENT, and Papua, end-users experience more frequent and longer duration outages compared to the reported SAIDI and SAIFI. Users experience a larger number of outages and longer duration for each interruption than those that are reported by the authority. P-SAIFIs are 4-fold to 14-fold higher than the PLN's SAIFIs. Also, P-SAIDIs are 8–12 times the PLN's SAIDIs for the corresponding provinces. As far as we know, this is the first independent study in Indonesia to evaluate the user experiences on the reliability of the power supply by the distribution grid and how the user experiences compare to the reported data from the utility. It can be concluded that the reliability of the power supply in these three cities in Indonesia could be improved considerably.

Acknowledgements We gratefully acknowledge the financial support provided for this project by the Indonesia Endowment Fund for Education (LPDP). Also, we highly appreciate the expert support provided by the Advanced Research on Innovations in Sustainability and Energy (ARISE) of the University of Twente (The Netherlands) and constructive comments by the editors and anonymous reviewers. Data were supported by PLN offices in Riau, ENT, and Papua. Measurements were accommodated by the Faculty of Science and Technology of UIN Suska Riau University in Pekanbaru and BMKG Offices in Kupang and Jayapura. Questionnaires were delivered with access to respondents arranged by students at the Department of Electrical Engineering at UIN Suska Riau University in Pekanbaru, Kumar Saputra and Ansel Mustalo in Kupang, and Jusuf Haurissa and Suyatno of the USTJ in Jayapura.

References

1. International Energy Agency, *World Energy Outlook 2016 Executive Summary* (International Energy Agency, Paris, 2016)
2. PLN, *Statistik PLN 2015* (PLN, Jakarta, 2016)
3. J. Khoury, R. Mbayed, E. Salloum, E. Monmasson, Optimal sizing of a residential PV-battery backup for an intermittent primary energy source under realistic constraints. Energy Build. **105**, 206–216 (2015)
4. M. Alramlawi, A. Gabash, P. Li, Optimal operation strategy of a hybrid PV-battery system under grid scheduled blackouts, in *IEEE International Conference on Environment and Electrical Engineering and 2017 IEEE Industrial and Commercial Power Systems Europe (EEEIC/I&CPS Europe)* (2017)
5. M.H. Hasan, T.M.I. Mahlia, H. Nur, A review on energy scenario and sustainable energy in Indonesia. Renew. Sustain. Energy Rev. **16**, 2316–2328 (2012)

6. P.M. Murphy, S. Twaha, I.S. Murphy, Analysis of the cost of reliable electricity: a new method for analyzing grid connected solar, diesel and hybrid distributed electricity systems considering an unreliable electric grid, with examples in Uganda. Energy **66**, 523e534 (2014)
7. World Bank Group, Enterprise surveys data for Indonesia, vol. 2017. Enterprise Analysis Unit, Global Indicators Department, DEC, World Bank, Washington DC, p. This page summarizes Enterprise Surveys data for I (2017)
8. IEEE, *IEEE Guide for Electric Power Distribution Reliability Indices*, vol. IEEE Std 1 (IEEE, New York, 2012)
9. Government of Indonesia, *National Energy General Plan 2016–2050*, vol. 22/2017 (Government of Indonesia, Jakarta, 2017)
10. D. Dalet, Indonesia. D-Maps (2007)
11. Asian Development Bank, *Achieving Universal Electricity Access in Indonesia* (Asian Development Bank, Philippines, 2016), p. 113,
12. The World Bank, *Acess to Electricity* (The World Bank, 2017)
13. International Agency Agency, *Electricity Access in Developing Asia—2016* (International Energy Agency, Paris, 2016)
14. Knoema, *Quality of electricity supply* (The World Bank, 2014)
15. CRO Forum, *Power Blackout Risks: Risk Management Options* (2011)
16. Q.F. Erahman, W.W. Purwanto, M. Sudibandriyo, A. Hidayatno, An assessment of Indonesia's energy security index and comparison with seventy countries. *Energy* 364–376 (2016)
17. Z.H. Jamal Hisham Hashim, Climate change, extreme weather events, and human health implications in the asia pacific region. Asia Pacific J. Public Heal. **28**(2S), 8S–14S (2016)
18. R. Zoro, R. Mefiardhi, Insulator damages due to lightning strikes in power system : some experiences in Indonesia, in *2006 IEEE 8th International Conference on Properties & applications of Dielectric Materials* (2006)
19. H.J. Christian et al., Global frequency and distribution of lightning as observed from space by the optical transient detector. J. Geophys. Res. Atmos. **108**(D1), ACL 4–1-ACL 4-15 (2003)
20. C. Bi, B. Qi, Reliability improvement of long distance transmission line protection in Indonesia, in *2016 China International Conference on Electricity Distribution (CICED)* (IEEE, Xi'an, 2016)
21. NASA, *Lightning Flash Rate*
22. S. Chattopadhyay M. Mitra, S. Sengupta, *Electric Power Quality* (Springer, Dordrecht, 2011)
23. J. Pontt, J. Rodríguez, W. Valderrama, G. Sepúlveda, G. Alzamora, Resonance effects, power quality and reliability issues of high-power converters-fed drives employed in modern SAG circuits. Miner. Eng. **17**(11–12), 1113–1125 (2004)
24. L. Sjöberg, Political decisions and public risk perception. Reliab. Eng. Syst. Saf. **72**(2), 115–123 (2001)
25. A. Ozbafli, G.P. Jenkins, Estimating the willingness to pay for reliable electricity supply: a choice experiment study. Energy Econ. **56**, 443–452 (2016)
26. R. Jimenez, T. Serebrisky, J. Mercado, What does 'better' mean? Perceptions of electricity and water services in Santo Domingo. Util. Policy **41**, 15–21 (2016)
27. A. Bernués, T. Rodríguez-Ortega, R. Ripoll-Bosch, F. Alfnes, Socio-cultural and economic valuation of ecosystem services provided by mediterranean mountain agroecosystems. PLoS One **9**(7) (2014)
28. B. O'Connor, R. Balasubramanyan, B.R. Routledge, N.A. Smith, From tweets to polls: linking text sentiment to public opinion time series, in *Fourth International AAAI Conference on Weblogs and Social Media* (2010)
29. I. Matsukawa, Y. Fujii, Customer preferences for reliable power supply: using data on actual choices of back-up equipment. Rev. Econ. Stat. **76**(3), 434–446 (1994)
30. J. Sagebiela, K. Rommel, Preferences for electricity supply attributes in emerging megacities—policy implications from a discrete choice experiment of private households in Hyderabad, India. Energy. Sustain. Dev. **21**, 89–99 (2014)
31. J. Sagebiel, J.R. Müller, J. Rommel, Are consumers willing to pay more for electricity from cooperatives? Results from an online choice experiment in Germany. Energy Res. Soc. Sci. **2**, 90–101 (2014)

32. B.J. Kalkbrenner, K. Yonezawa, J. Roosen, Consumer preferences for electricity tariffs: does proximity matter? Energy Policy **107**, 413–424 (2017)
33. A. Bartczak, S. Chilton, M. Czajkowskia, J. Meyerhoff, Gain and loss of money in a choice experiment. The impact of financial loss aversion and risk preferences on willingness to pay to avoid renewable energy externalities. Energy Econ. **65**, 326–334 (2017)
34. J. Sagebiel, Preference heterogeneity in energy discrete choice experiments: A review on methods for model selection. Renew. Sustain. Energy Rev. **69**, 804–811 (2017)
35. J. Shina, W.-S. Hwang, Consumer preference and willingness to pay for a renewable fuel standard (RFS) policy: focusing on ex-ante market analysis and segmentation. Energy Policy **106**, 32–40 (2017)
36. H. Harapan et al., Willingness to pay for a dengue vaccine and its associated determinants in Indonesia: a community-based, cross-sectional survey in Aceh. Acta Trop. **166**, 249–256 (2017)
37. D. Vollmer, A.N. Ryffel, K. Djaja, A. Grêt-Regamey, Examining demand for urban river reha-bilitation in Indonesia: insights from a spatially explicit discrete choice experiment. Land Use Policy **57**, 514–525
38. Y. Suparman, H. Folmer, J.H.L. Oud, The willingness to pay for in-house piped water in urban and rural Indonesia. Pap. Reg. Sci. **95**(2), 407–426
39. A. Ghozali, S. Kaneko, *Climate Change Policies and Challenges in Indonesia Consumer behavior and ecolabeling* (Springer, Japan, 2016)
40. J.S.T. Soo, Valuing air quality in Indonesia using households' locational choices. Environ. Resour. Econ. 1–22 (2017)
41. Z.Z.Z. Anna, Economic valuation of whale shark tourism in Cenderawasih Bay National Park, Papua, Indonesia. Biodiversitas **18**(3), 1026–1034 (2017)
42. P. Hendratmoko, Guritnaningsih, T. Tjahjono, Analysis of interaction between preferences and intention for determining the behavior of vehicle maintenance pay as a basis for transportation road safety assessment, vol. 7, no. 1, (2016), pp. 105–113
43. L. Ambarwati, R. Verhaeghe, B.V. Arem, A.J. Pel, Assessment of transport performance index for urban transport development strategies—incorporating residents' preferences. Environ. Impact Assess. Rev. **6**, 107–115 (2017)
44. R. Kojima, M. Ishikawa, Consumer willingness-to-pay for packaging and contents in Asian countries. Waste Manag. (2017)
45. S. Miller, P. Tait, C. Saunders, P. Dalziel, P. Rutherford, W. Abell, Estimation of consumer willingness-to-pay for social responsibility in fruit and vegetable products: A cross-country comparison using a choice experiment. J. Consum. Behav. **16**(6), e13–e25 (2017)
46. R. Lensink, T. Raster, A. Timmer, Liquidity constraints and willingness to pay for solar lamps and water filters in Jakarta. Eur. J. Dev. Res. (2017)
47. H. Kumashiro, V. Kharisma, C.P. Morgana Sianipar, K. Koido, R. Takahashi, K. Dowaki, The design of an appropriate geothermal energy system, in *5th International Conference on Energy and Sustainability* (WITPress, Putrajaya, Malaysia, 2014), pp. 207–217
48. PLN, *Statistik PLN 2010* (PLN, Jakarta, 2011)
49. PLN, *Statistik PLN 2011* (PLN, Jakarta, 2012)
50. PLN, *Statistik PLN 2012* (PLN, Jakarta, 2013)
51. PLN, *Statistik PLN 2013*, no. 02601 (PLN, Jakarta, 2014)
52. PLN, *Statistik PLN 2014* (PLN, Jakarta 2015)
53. B. Matthews, L. Ross, *Research Methods: A Practical Guide for the Social Sciences* (Pearson Longman, 2010)
54. J.A. List, C.A. Gallet, What experimental protocol influence disparities between actual and hypothetical stated values? Environ. Resour. Econ. **20**(3), 241–254 (2001)
55. S. Manikandan, Measures of central tendency: the mean. J. Pharmacol. Pharmacother. **2**(2), 140–142 (2011)
56. Mathcentre, Variance and standard deviation (grouped data) (Mathcentre, Loughborough, 2003)

57. Ministry of Energy and Mineral Resources, *Minister of Energy and Mineral Resources Decree No. 18/2016 regarding Sumatra Electric Power System Regulation* (Ministry of Energy and Mineral Resources, Jakarta, 2016)
58. J.E. DeCaria, M. Montero-Odasso, D. Wolfe, B.M. Chesworth, R.J. Petrella, The effect of intra-articular hyaluronic acid treatment on gait velocity in older knee osteoarthritis patients: a randomized, controlled study. Arch. Gerontol. Geriatr. **55**(2), 310–315 (2012)
59. V. Štěrbová, J. Kupka, J. Thomas, J. Lichnovský, P. Andráš, Land snail assemblages of production forest in relation to selected environmental factors (HrabĚtice forest, Czech republic), in *International Multidisciplinary Scientific Geo Conference on Surveying Geology and Mining Ecology Management, SGEM*, vol. 2 (2015), pp. 295–299
60. W.A. Stout Jr., B. Tawney, An excel forecasting model to aid in decision making that affects hospital resource/bed utilization-hospital capability to admit emergency room patients," in *Proceedings of the 2005 IEEE Systems and Information Engineering Design Symposium* (2005), pp. 222–228
61. F. Carlsson, P. Martinsson, Willingness to pay among swedish households to avoid power outages: a random parameter tobit model approach. Energy J. **28**(1), 75–89 (2007)
62. C. Kirubi, A. Jacobson, D.M. Kammen, A. Mills, Community-based electric micro-grids can contribute to rural development: evidence from Kenya. World Dev **37**(7), 1208–1221
63. L. Goel, A comparison of distribution system reliability indices for different operating configurations. Electr. Mach. Power Syst. **27**(9), 1029–1039 (1999)

Chapter 4
The Attitudes of End-Users Toward Solar Photovoltaics

4.1 Preface

In Chap. 3, we presented the results from a series of surveys at three different locations in Indonesia, namely in the islands of Sumatra, Timor, and Papua. The discussion focused on the experiences of end-users of the electricity grid regarding the reliability of the electricity supply. In this chapter, using data from the same survey, we will present additional findings regarding the awareness of the households about renewable energy and climate change and their attitudes toward solar photovoltaics. The methods used in data collection and data analysis remain the same as the method used in Chap. 3 [1]. As such, for an explanation of the research set-up, refer to Sect. 3.3.

4.2 The Survey

Data were collected through a structured questionnaire using closed-ended questions in combination with 'face-to-face' semi-structured interviews. To capture the opinions and attitudes of end-users, the questionnaire contained ten questions covering various topics regarding renewable energy, climate change, and PV systems (see Appendix A.3, Table A.3).

1. Have you heard of 'renewable energy'?
2. Is renewable energy important for Indonesia?
3. Have you heard of 'climate change'?
4. Are you worried about climate change?
5. Have you heard of 'PV systems'?
6. Which one of the following two electricity sources do you believe is cheaper; the grid or a PV system?

© The Author(s), under exclusive license to Springer Nature Switzerland AG 2020   67
K. Kunaifi et al., *The Electricity Grid in Indonesia*,
SpringerBriefs in Applied Sciences and Technology,
https://doi.org/10.1007/978-3-030-38342-8_4

7. Which one of the following two electricity sources do you believe is better for the environment; the grid or a PV system?
8. Which one of the following two electricity sources do you believe is more stable; the grid or a PV system?
9. Which one of the following two electricity sources would you choose to provide power supply to your home; the grid or a PV system?
10. Would you like to have a PV system installed on your house's rooftop?

4.3   Awareness of End-Users About Renewable Energy and Climate Change

The first set of questions on the survey considered the awareness of the participants about renewable energy and climate change. The participants were using power from the grid at the time of the surveys. They expressed their knowledge on renewable energy and climate change and their opinion about the importance of renewable energy. They also showed whether or not they were worried about climate change.

It could be concluded that most of the participants were knowledgeable about renewable energy. This is represented by more than half of the participants in Pekanbaru (55%), who acknowledged that they knew something about renewable energy, and even higher fractions in Kupang and Jayapura, with values of 60 and 77%, respectively (Fig. 4.1a and Appendix A.3 (Table A.3)). Based on this response, it can be expected that most of the participants acknowledged the importance of renewable energy. As shown in Fig. 4.1b and Appendix A.3 (Table A.3), 51, 57, and 81% of the participants in Pekanbaru, Kupang, and Jayapura, respectively, believe that renewable energy is important for Indonesia. Figure 4.1a, b shows that the level of knowledge of the participants on renewable energy influences their belief about the importance of renewable energy. The more they know about renewable energy, the more important they perceived renewable energy to be.

The results from the survey also showed that most of the participants were knowledgeable about climate change. Sixty-nine percent of the participants in Pekanbaru responded positively, while in Kupang and Jayapura, 58 and 92% of the participants confirmed this as well (Fig. 4.1c and Appendix A.3 (Table A.3)). Their knowledge about climate change seemed to influence their attitude toward climate change. As shown in Fig. 4.1d and Appendix A.3 (Table A.3), the majority of participants were worried about climate change; 60% in Pekanbaru, 51% in Kupang, and 85% in Jayapura. As such, it can be concluded that awareness and concerns about climate change are closely related.

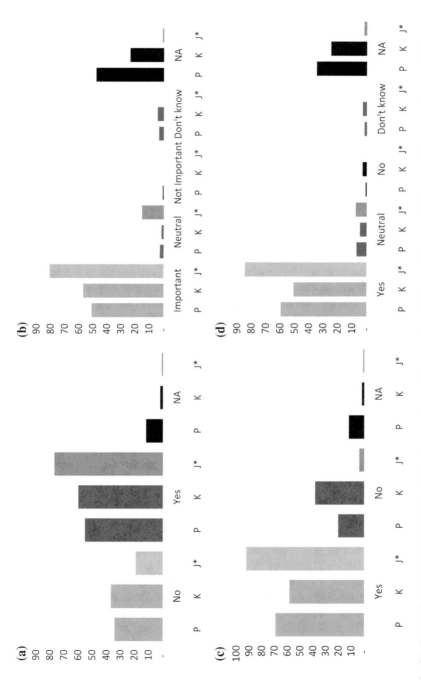

Fig. 4.1 Respondents' answers to survey questions in percentages: **a** Have you heard of 'renewable energy'? **b** Is renewable energy important for Indonesia? **c** Have you heard of 'climate change'? **d** Are you worried about climate change? P: Pekanbaru, K: Kupang, J: Jayapura, J: Jayapura *, NA: No answer. * Data from Jayapura is less representative

4.4   Attitudes of the End-Users of the Electricity Grid Toward PV Systems

The second set of questions on the survey considered the awareness and attitudes of the participants with regards to PV systems. First, they expressed their knowledge about PV systems in terms of the cost, sustainability, and reliability of PV systems compared to the electricity from the grid. For this purpose, the participants answered questions 5–8 (see Sect. 4.2). Next, they showed their attitude toward PV systems by responding to questions 9–10 about their preference for the electricity sources at home, either the grid or PV systems, and their willingness to have a PV array installed on their house's rooftop.

Figure 4.2 shows interesting results. Namely, most of the respondents in the three cities had some knowledge about PV systems. This is similar to their response to the previous question regarding their knowledge about renewable energy, except that in Kupang, a lower share of the participants knew about PV systems compared to those who knew about renewable energy.

However, all participants had limited knowledge about the difference in the price of electricity from the grid and PV systems. Most of the participants in Pekanbaru did not know which one is cheaper, while a significant number of participants in Pekanbaru believed that PV systems generate more affordable energy than the grid does. Differently, in Kupang and Jayapura, the majority of the participants believed that the price of electricity from PV systems is lower than the electricity from the grid (PLN) (Fig. 4.2b and Appendix A.3 (Table A.3)). The lack of knowledge of the participants about the difference in price between the electricity from the grid and PV systems is logical because none of the participants had experience with PV systems. In fact, this is a difficult question, even for an expert.

The truth is that under the official tariffs, the price of electricity from PV systems was more than two times higher than the price of electricity from the grid. However, the price of electricity from the grid paid by the end-users did not reflect the true price of energy production because the government subsidizes the price. Without a subsidy for the electricity from the grid, the price of PV power is just a bit higher than the price of electricity from the grid. In the near future, the price of PV energy is likely to be the same as, or even lower, than the price of energy from the grid.

Most of the participants in the three cities knew that PV systems generate more sustainable electricity as compared to the electricity from the grid providing the current practice of power generation in Indonesia (Fig. 4.2c and Appendix A.3 (Table A.3)). However, most of the participants did not know which power source is more reliable: PV systems or the grid? In Pekanbaru and Kupang, the majority of the participants believed that the power from the grid is more stable, while in Kupang, the participants were more favorable toward PV systems in terms of the reliability of electricity supply. This confirms the findings of our study in Chap. 3, where in Kupang, the participants experienced worse power outages by the grid compared to those in Pekanbaru and Jayapura.

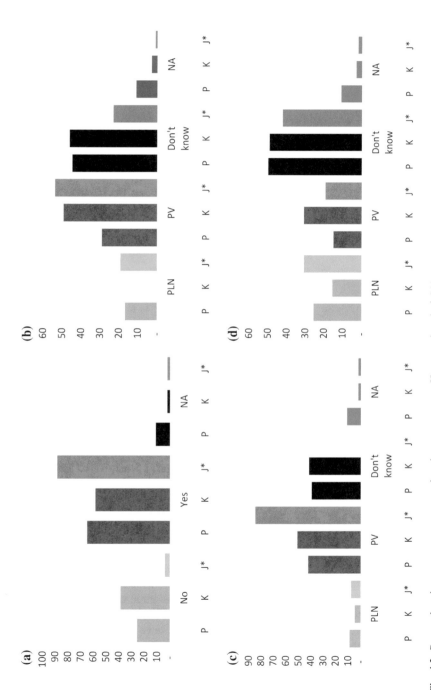

Fig. 4.2 Respondents' answers to survey questions in percentages: **a** Have you heard of 'PV systems'? **b** Which one of the following two electricity sources do you believe is cheaper? **c** Which one of the following two electricity sources do you believe is better for the environment? **d** Which one of the following two electricity sources do you believe is more stable? P: Pekanbaru, K: Kupang, J: Jayapura *, NA: No answer. * Data from Jayapura is less representative

Fig. 4.3 Respondents' answers to survey questions in percentages: **a** Which one of the following two electricity sources would you choose to provide power supply to your home? **b** Would you like to have a PV system installed on your house's rooftop? P: Pekanbaru, K: Kupang, J: Jayapura *, NA: No answer. * Data from Jayapura is less representative

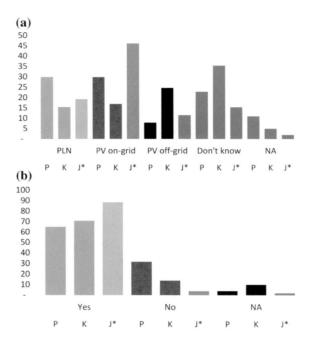

Figure 4.3 portrays the attitude of the end-users of the electricity grid in the three cities in Indonesia toward PV systems. First, the participants were asked a question, 'which of the following electricity sources would you choose for your house; the grid (PLN), on-grid PV systems, or off-grid PV systems?' In Pekanbaru, the percentage of participants who favored the grid and on-grid PV systems was the same. A lower share of the participants in Pekanbaru chose off-grid PV systems. However, a significant proportion of participants in Pekanbaru did not know what to choose from the three options.

In Kupang, while the majority of the participants did not know the best option for them, the second largest percentage of participants favored off-grid PV systems, followed by those who wanted on-grid PV systems, and the lowest percentage chose electricity from the grid.

In Jayapura, 58% of the participants favored PV systems, where most of them wanted on-grid installations. Only 19% of the participants in Jayapura selected the grid, while 15% of the participants did not know the answer to the question.

Finally, the participants were asked, 'do you want a PV array to be installed on your house's rooftop?' Most of the participants in the three cities favored PV systems on their house's rooftops; 65% in Pekanbaru, 71% in Kupang, and 88% in Jayapura. The majority of the participants who refused to have rooftop PV systems were concerned about the strength of their houses' roofs for accommodating PV system components and structures.

4.5 Conclusions

The number of participants in Pekanbaru and Kupang was higher than in Jayapura. Therefore, for this study, data from Jayapura is less representative, although it is still beneficial given the difficulties of getting data from Jayapura. Also, regarding the level of education, participants in Pekanbaru and Kupang were more varied, while in Jayapura, the participants who participated in this survey were much more educated than average.

The responses and preferences from the participants in this study were strongly influenced by their levels of education and experience. Participants in Jayapura, who had higher education than those in Pekanbaru and Kupang, knew and supported renewable energy more. Most of them knew about climate change, and they worried about that. They had a positive attitude toward PV systems regarding energy price, sustainability, and reliability. Finally, they chose PV systems and wanted to have one at home.

The participants from Pekanbaru and Kupang were more diverse in terms of education and economic condition compared to participants in Jayapura. Therefore, the responses from participants in Pekanbaru and Kupang were highly influenced by their experience with the grid. Participants in Kupang experienced less reliable electricity supply compared to those in Pekanbaru. Thus, they favored PV systems more than participants in Pekanbaru. Even participants in Kupang wanted to have off-grid PV systems, which means to become independent from the grid (PLN).

Despite the variety in responses to the questions in the questionnaires from the participants, many households in Indonesia wanted to have PV systems on their houses' rooftops, either to increase the reliability of electricity supply as the user of the grid or to become separated from the grid.

Reference

1. A. Kunaifi, A. Reinders, Perceived and reported reliability of the electricity supply at three urban locations in Indonesia. Energies (2018)

Chapter 5
The Energy Potential of Solar Photovoltaics in Indonesia

5.1 Preface

A few years ago, we developed a method for mapping the potential of grid-connected and off-grid PV systems [1, 2] in Indonesia[1]. In this chapter, we will apply a slightly modified version of the same method with more recent data from the year 2018. As such, a more up-to-date quantitative estimate of the technical potential and cost of PV systems in Indonesia will be calculated with the purpose of gaining better insight into the actual and maximally achievable installation capacity of PV systems in various parts of Indonesia. This chapter will, therefore, be structured as follows. The detailed methodology for calculation of the potential and costs of grid-connected and off-grid PV systems will be presented in Sect. 5.2. The calculation results will be presented in Sect. 5.3 for the potential of grid-connected PV systems and in Sect. 5.4 for the potential of off-grid PV systems. An overview of the full calculations will be shown in Sect. 5.5, followed by discussion in Sect. 5.6 and conclusions in Sect. 5.7.

5.2 Methodology

The potential of PV systems in Indonesia will be modeled using publicly available data. Since large geographical differences exist in Indonesia and most of the relevant

---

[1]This chapter is based on A. J. Veldhuis and A. H. M. E. Reinders, "Reviewing the potential and cost-effectiveness of grid-connected solar PV in Indonesia on a provincial level," Renew. Sustain. Energy Rev., vol. 27, pp. 315–324, 2013 [1], and A. J. Veldhuis and A. H. M. E. Reinders, "Reviewing the potential and cost-effectiveness of off-grid PV systems in Indonesia on a provincial level," Renew. Sustain. Energy Rev., vol. 52, pp. 757–769, Aug. 2015 [2].

data are available on a provincial level, this study examines the potential of off-grid PV systems for each of the 33 provinces.[2]

In this study, we distinguish off-grid and grid-connected systems. To determine the potential of both systems a number of assumptions have to be made. First, it is assumed that the electricity supply follows electricity demand. This implies that electricity consumption in a certain province is an important input variable. Second, it is assumed that if at a specific location an electricity grid is available, grid-connected PV systems will be favored over off-grid PV systems, though off-grid or autonomous systems can co-exist with grid-connected PV systems. In cases where a location lacks an electricity grid, off-grid PV systems will be able to meet the electricity demand in our model.

Since the approach for grid-connected PV systems differs from the approach for off-grid PV systems, the methodologies for both systems will be presented in separate sections below. Off-grid systems cannot rely on the electricity grid for power supply when the PV systems deliver no or not enough power, so off-grid systems either need a diesel generator or an energy storage system to guarantee sufficient energy availability throughout the year. Since batteries are still relatively expensive, this study assumes a hybrid diesel-PV-battery system for rural villages with adequate demand and stand-alone systems for smaller settlements. This difference in the method to determine the potential of PV systems in Indonesia for grid-connected and off-grid systems does not reduce the usefulness of the results, since the potential of grid-connected systems is zero in areas without an electricity grid, and with current technologies, PV system penetration rates and costs, grid-connected PV is more cost-effective than hybrid or stand-alone systems in cases where an electricity grid is present. Therefore, it has been assumed that grid-connected and off-grid PV systems complement each other.

5.2.1   Grid-Connected PV Systems

5.2.1.1   Suitable Areas

For grid-connected PV systems, the presence of a grid is a prerequisite to be able to feed electricity into the grid. Since data about geographic locations with a grid connection are not publicly available in Indonesia, a method has been developed to determine the areas suitable for grid-connected PV systems based on the available data of land area, population, electrification ratio and urbanization ratio per province [1].

---

[2]In 2012 the northern part of the province of East Kalimantan has formed its own province: North Kalimantan. However, most publicly available data is still only available for the former province of East Kalimantan, therefore we will use the data of the former province for the whole area, including North Kalimantan as well.

The PLN statistics of 2018 show the electrification ratio per province [3], and the values have been plotted in Fig. 5.1. From the figure it can be seen that the electrification ratio varies between 42% in the province of Papua to 100% in several other provinces. This is better than the situation a few years ago in 2010. Namely, in our previous study, the electrification ratio varied from 24 to 93%.

To calculate the potential of grid-connected PV, the area suitable for grid-connected PV must be determined. Since an electricity grid is expensive and only cost-effective in densely populated areas, we expect that these areas have priority over sparsely populated areas regarding grid connection. To model the suitable area, in this study, each province has been divided into four different areas ordered from high to low population density: (1) urban cores, (2) suburbs, (3) villages and (4) rural areas. It is assumed that the electricity grid is extended in the same sequence, from urban cores to rural areas. Furthermore, it is assumed that rural areas do not have an electricity grid and are not suitable for grid-connected PV.

To determine the share of the population which is connected to the grid, we assume that the household size is homogeneous inside each province, so that the electrification ratio in a province ER_p can be multiplied by the population in the province N_p (both data from PLN [3]) in order to obtain the population with grid access:

$$N_{p,\text{grid}} = N_p \times ER_p \tag{5.1}$$

The population in the urban core with grid connection in province p, $N_{p,\text{grid},1}$, is determined as follows:

$$N_{p,\text{grid},1} = \begin{cases} N_{p,1}, N_{p,1} \leq N_{p,\text{grid}} \\ N_{p,\text{grid}}, N_{p,1} > N_{p,\text{grid}} \end{cases} \tag{5.2}$$

where $N_{p,1}$ is the population living in the urban core according to the Bureau of Statistics (BPS). The most recent data is from the population census from 2010 [4], which shows the urban and rural population data in municipalities within a province. BPS also published data on the population projection for each province per five years from 2010 to 2035 [5]. The provincial growth rate from 2010 to 2020 has been used to account for the population growth in urban cores. It has been assumed that the population growth per province is uniformly distributed over the various areas within a province. The Indonesian population is expected to grow by 14% during this decade; however, per province this varies between 6 and 32%. The total population in the suburbs and urban cores of a province, $N_{p,1,2}$, is determined as follows:

$$N_{p,1,2} = N_p \times UR_p \tag{5.3}$$

where UR_p is the urbanization ratio of the province according to BPS for the year 2020 [6]. They classify a village as urban if it satisfies the following three conditions: (a) a population density of at least 5,000 persons per square kilometer, (b) 25% or less of

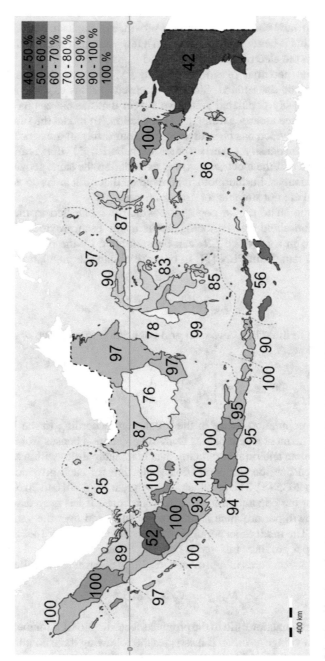

Fig. 5.1 Electrification ratio of households per province in Indonesia in 2018 [3]

the households working in the agriculture sector, and (c) eight or more urban-related facilities like post office, bank, cinema, hospital, and school.

Next, the suburban population connected to the grid, $N_{p,\text{grid},2}$, is determined according to:

$$N_{p,\text{grid},2} = \begin{cases} N_{p,1,2} - N_{p,\text{grid},1}, & N_{p,1,2} \leq N_{p,\text{grid}} \\ N_{p,\text{grid}} - N_{p,\text{grid},1}, & N_{p,1,2} > N_{p,\text{grid}} \end{cases} \tag{5.4}$$

Similarly, the population in villages with grid-connection, $N_{p,\text{grid},3}$, is determined:

$$N_{p,\text{grid},3} = \begin{cases} N_{p,\text{grid}} - N_{p,1,2}, & N_{p,1,2} < N_{p,\text{grid}} \\ 0, & N_{p,1,2} \geq N_{p,\text{grid}} \end{cases} \tag{5.5}$$

Now that the populations for the various areas can be determined, the corresponding land areas can be calculated. The sizes of the various areas are obtained by dividing the population of the distinguished area by its average population density. For simplicity, the average population density in Indonesia for these distinguished areas is assumed equal for each province; see Table 5.1.

The land area of the urban core with grid-connection is calculated as follows:

$$A_{p,\text{grid},1} = \begin{cases} \frac{N_{p,\text{grid},1}}{\overline{ND}_1}, & \frac{N_{p,\text{grid},1}}{\overline{ND}_1} \leq A_{p,0.8} \\ A_{p,0.8}, & \frac{N_{p,\text{grid},1}}{\overline{ND}_1} > A_{p,0.8} \end{cases} \tag{5.6}$$

where $A_{p,\text{grid},1}$ is the provincial urban core area in km^2 wherein the population is connected to the electricity grid, $\overline{ND}_1$ is the average population density of urban cores in Indonesia in persons/km^2 and $A_{p,0.8}$ is 80% of the total land area of the province in km^2. Assumed is that at least 20% of the land area of each province is not populated, which affects the island of Java only, because of the high population density. For the Special Capital Region of Jakarta, an exception has been made, and 100% of the total land area can be populated since it is a relatively small area (664 km^2) and very densely populated.

The suburban area is determined in a similar way:

$$A_{p,\text{grid},2} = \begin{cases} \frac{N_{p,\text{grid},2}}{\overline{ND}_2}, & \frac{N_{p,\text{grid},2}}{\overline{ND}_2} \leq A_{p,0.8} \\ A_{p,0.8} - A_{p,\text{grid},1}, & \frac{N_{p,\text{grid},1}}{\overline{ND}_2} > A_{p,0.8} \end{cases} \tag{5.7}$$

The area of the grid-connected villages can be found using the following formula:

$$A_{p,\text{grid},3} = \begin{cases} \frac{N_{p,\text{grid},3}}{\overline{ND}_3}, & \frac{N_{p,\text{grid},3}}{\overline{ND}_3} \leq A_{p,0.8} \\ A_{p,0.8} - (A_{p,\text{grid},1} + A_{p,\text{grid},2}), & \frac{N_{p,\text{grid},3}}{\overline{ND}_3} > A_{p,0.8} \end{cases} \tag{5.8}$$

where $A_{p,\text{grid},3}$ is the area in km^2 of the villages in the province wherein the population is connected to the electricity grid and $\overline{ND}_3$ is the average population density of villages in Indonesia.

By assuming a constant population density for each of the specific areas, the area of the population with a grid connection can be found. The population density of 8,000 persons per square kilometer for urban cores is based on the average population density of the largest cities in Indonesia. The population density of 5,000 persons per square kilometer for suburbs is the lowest for an area to be classified as urban according to BPS. The population density of 1,000 residents per square kilometer is assumed to be the average for rural villages with grid-connection in Indonesia.

The technical production potential of grid-connected PV systems for each province of Indonesia is calculated using the following formula:

$$E_{p,\text{PV,pot}} = d_y . \eta . H_p . \sum_{i=1}^{3} LA_i . A_{p,\text{grid},i} . PR_i \qquad (5.9)$$

where $E_{\text{PV},p,\text{pot}}$ is the potential annual electricity production of grid-connected PV of the province in GWh/year, d_y is the number of days per year, η is the efficiency of the PV modules, PR is the performance ratio, H_p is the average irradiation in kWh/m^2/d of the province, i is the area type (1 = urban core, 2 = suburban, 3 = village) and LA is the land availability factor for grid-connected PV. In Table 5.1 the values for the various factors are presented.

In this study, the selected PV technology is multi-crystalline silicon. The module efficiency (η) of multi-crystalline silicon is assumed to be 17.3% (assuming a 285 W_p module). The performance ratio (PR) of grid-connected PV in Indonesia is estimated to be 75% for urban cores and 80% for suburban areas. These values are in line with European systems. Because of higher temperatures and shading due to surroundings, the PR for urban cores is assumed to be 5% lower compared with suburban areas. The PR of grid-connected PV is assumed to be 70% in rural areas, due to lower and more unstable electricity demand, increased soiling of PV modules in rural areas,

Table 5.1 Assumptions for some parameters for calculation of the potential of grid-connected PV

Variable	Value	Unit
$\overline{ND}_1$	8,000	Residents/km^2
$\overline{ND}_2$	5,000	Residents/km^2
$\overline{ND}_3$	1,000	Residents/km^2
LA_1	5	%
LA_2	10	%
LA_3	15	%
PR_1	75	%
PR_2	80	%
PR_3	70	%

which are known to be more 'dusty,' and higher transmission losses due to longer distances.

5.2.1.2 Electricity Demand

The total amount of electricity generated by grid-connected PV should not exceed the electricity demand in a particular province. The electricity demand per province is obtained from the statistics of PLN [3], which contains information about the amount of energy sold per customer type (GWh). Since no energy storage system is assumed, per customer type an assumption has been made regarding the electricity demand during Indonesian daylight time. Further, it is assumed that this demand pattern is uniform across Indonesia (Table 5.2).

An extra limitation of the total electricity produced by PV is related to the flexibility of existing base load power plants. Conventional power generators often require a minimum power output and cannot easily be turned completely off and on again. Therefore, we assume a minimum base load supplied by conventional generators per province. It is assumed that the base load equals 40% of the electricity demand during night-time and that the minimal load constraint is assumed to be 90% of this amount:

$$E_{p,\text{load,min}} = E_{p,\text{demand,night}} \times 0.4 \times 0.9 \tag{5.10}$$

where $E_{p,\text{load,min}}$ is the minimal electricity produced by existing electric power systems and $E_{p,\text{demand,night}}$ is the electricity demand during night-time of the province, both in GWh/year.

The maximum amount of electricity generated by PV $E_{p,\text{PV,demand}}$, based on the electricity demand is therefore:

$$E_{p,\text{PV,demand}} = E_{p,\text{demand,day}} - E_{p,\text{load,min}} \tag{5.11}$$

where $E_{p,\text{demand,day}}$ is the electricity demand of the province during daylight time.

Table 5.2 Assumptions of the electricity demand during daylight time for each customer type of PLN	Customer type	Demand daylight time (%)	Demand night time (%)
	Residential	40	60
	Industrial	80	20
	Business	60	40
	Social	70	30
	Governmental building	90	10
	Street lighting	0	100

The yearly amount of electricity that can be supplied by grid-connected PV in a province, $E_{p,\text{PV}}$, is:

$$E_{p,\text{PV}} = \min(E_{p,\text{PV,demand}}, E_{p,\text{PV,pot}}) \tag{5.12}$$

The produced electricity by PV per capita can be obtained by:

$$E_{p,\text{PV,capita}} = \frac{E_{p,\text{PV}}}{N_{p,\text{grid}}} \tag{5.13}$$

Assuming identical electricity consumption per capita in each province, the electricity generated by PV per area type (e.g. urban) can be determined by multiplying the $E_{p,\text{PV,capita}}$ by the population connected to the grid in this area.

$$E_{p,\text{PV},i} = E_{p,\text{PV,capita}} \times N_{p,\text{grid},i} \tag{5.14}$$

In Appendix B.1, a large share of the input data is shown for each province, including: land area (A), population (N), urbanization ratio (UR), population in urban cores ($N_{\text{urban\_core}}$), electrification ratio (ER) and irradiation (H).

5.2.2   Off-Grid PV Systems

To determine the potential of off-grid PV systems in Indonesia, the approach is different. It is assumed that urban households which lack access to electricity will be on the national electricity grid's (PLN's) waiting list and will be connected to the grid in the near future, since this is the general policy of PLN. Besides, in these urban areas grid extension is assumed to be the most cost-effective electrification option, therefore the scope of this part of the study is limited to the rural households without access to electricity. As such, the province of Jakarta is excluded from this analysis, since it contains no rural households without electricity access at all, leaving only 32 provinces left for the off-grid part of the study.

To calculate the actual potential of off-grid PV systems the following general assumptions have been made:

(i)   Other renewable energy technologies such as hydropower, geothermal energy, wind energy or biomass are not taken into account.
(ii) Rural electricity demand is a dominant factor.
(iii) Off-grid PV systems are only viable in rural areas lacking grid-connection
(iv) The model is based on a provincial level, so regional variations inside the province (i.e. at a district level) itself are outside the scope of this study.
(v)   There are two PV system configurations evaluated in this study: (a) hybrid microgrid consisting of PV systems, batteries and diesel generators, and (b) stand-alone PV systems with batteries.

In order to determine the potential and costs of off-grid PV systems, the modeling approach is divided into several calculation steps. These steps are:

- Rural population without electricity access
- Off-grid electricity demand
- Rural population distribution
- PV system configuration
- PV system sizing.

In the following paragraphs each step is explained.

5.2.2.1 Rural Population Without Electricity Access

In order to obtain the number of rural households lacking electricity access, data from PLN from 2018 [3] and data from BPS [7] have been combined. BPS presents data for rural and urban households separately and they distinguish four categories: (1) PLN Electricity with Installation, (2) PLN Electricity without Installation, (3) Electricity Non-PLN and (4) Non Electricity. For this study, we are interested in the fourth category. Since more recent data are lacking, we assume that the proportion of households that have no electricity compared to the total number of households that have no PLN installation (that is the summation of the households in the categories 2-4) has remained the same over time.

In Fig. 5.2 the percentage of rural households that have no access to electricity is shown per province, ranging from 0 to 75%. Since the population distribution in Indonesia is heavily skewed, in Fig. 5.3 the distribution of those rural households across Indonesia is shown, which ranges from 0 to 27%.

In six provinces, all rural households are assumed to have access to electricity already. This accounts for the following provinces: Bangka Belitung, Central Java, North Sumatra, South Sumatra, West Java, and West Papua.

Overall, it is estimated that according to official reports of PLN, there are roughly 1.5 million households out of a total of 67.6 million households in all of Indonesia lacking electricity. The information obtained from both maps shows that the provinces of Papua, East Nusa Tenggara, and Jambi form the top three provinces where the largest share of rural households lacking electricity access can be found, both in relative (Fig. 5.2) and absolute numbers (Fig. 5.3).

5.2.2.2 Off-Grid Electricity Demand

Next, the rural electricity demand will be determined. Detailed data on rural electricity demand in Indonesia are lacking. The average household's electricity consumption in rural areas in developing countries varies between 240 and 768 kWh/year [8, 9]. Based on these figures in combination with the available data from PLN and BPS, it is assumed that an Indonesian household living off-grid would consume between

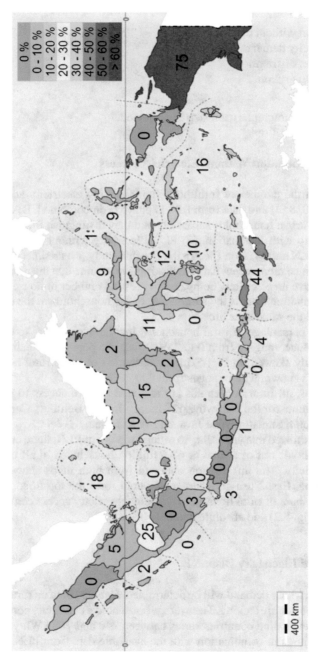

Fig. 5.2 Percentage of rural households lacking electricity access per province

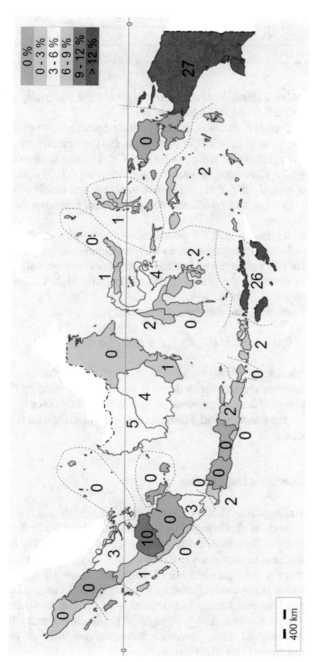

Fig. 5.3 Distribution of rural households lacking electricity access across Indonesia

Table 5.3 Assumed energy demand characteristics for the low and high demand categories

Category	E_{HH} (kWh/year)	L_{peak} (kW)	α
Low (mainly lighting, fan, TV & radio)	250	0.128	1.0
High (with productive use)	500	0.138	0.6

250 and 540 kWh/year. Further, it is assumed that the rural electricity demand is roughly the same for each province in Indonesia.

For the sizing of off-grid PV systems, the daily demand profile is important. Details are lacking; however, most households in rural areas use their electricity mainly for lighting and to a lesser extent for a fan, radio and TV. The electricity demand for lighting will largely take place between sunset and sunrise, especially during the time before sunrise from roughly 5–6 a.m., and between sunset and bedtime, so roughly from 6–10 p.m. In contrast, non-residential consumers will consume most of their electricity during daytime.

To be able to model the off-grid PV potential, two categories are proposed (see Table 5.1). The high demand category extends the low demand category. It is assumed that the electricity for productive use is required during working days only and mainly during the daytime, therefore the peak load is only slightly higher compared with the low demand category (Table 5.3).

The off-grid PV potential can then be determined by:

$$E_{p,\mathrm{PV}} = \alpha_h H H_{p,h} E_{HH,h} + \alpha_l H H_{p,l} E_{HH,l} \tag{5.15}$$

where $E_{p,\mathrm{PV}}$ is the electricity (kWh/year) generated by PV in province p, α ($0 \leq \alpha \leq 1$) determines the share of the electricity demand which is generated by PV, HH_p is the number of rural households in province p, E_{HH} is the yearly electricity demand of a household. The subscripts h and l indicate the high demand and low demand categories, respectively.

5.2.2.3 Rural Population Distribution

Now when the electricity demand of a rural household is modeled, it is necessary to obtain information about the clustering of these households in order to determine the energy demand per area. Therefore, the population density distribution of the rural population without access to electricity per province, $N_{p,rur,ne}$, should be known.

To be able to determine the rural electricity demand per area, we model the population density distribution in this study. Based on the assumption of average population densities (ND) for various areas made in the previous study, the population density distribution can be approximated. The geographical population density distribution often follows an exponential trend, therefore this study proposes an exponential function for the modeling of this distribution.

To calculate the actual population densities per province, some additional boundary conditions have to be set. Although most rural people live inside small settlements, it is assumed that in each province there is at least one area in which the population density is very low. This is modeled by the boundary condition that the lowest population density in a province is equal to or less than 10 persons/km² in populated areas. Further is assumed that the highest population density in non-electrified rural areas is smaller than the population density assumed for rural electrified villages, $ND_{rur,grid}$, which equals 1,000 persons/km². The rationale behind this is that if the population density were larger, the village would most likely already have been connected to the (local) grid, because of the favorable conditions for infrastructure investments in these areas. Based on these aforementioned assumptions, the rate of diminution in population density over distance, t_p, can be determined in accordance with the method described in [2].

Further it is assumed that each step i corresponds to a population density ND_i inside an area equal to 1 km², so that the population living in this area is numerically equivalent to the population density and i is numerically identical to the non-electrified rural area of province p, $A_{p,rur,ne}$, in km².

With the population distribution modeled, the electricity demand per area can be determined.

5.2.2.4 PV System Configuration

There are two PV system configurations evaluated in this study: (a) hybrid microgrid consisting of PV systems, batteries and diesel generators, and (b) stand-alone PV systems with batteries.

When a micro-grid is viable, it is assumed to be the most preferred option. According to a rule of thumb for the feasibility of isolated grids from SWECO [10], the number of connections c in an area within a radius r of 500 m should be at least 100. We assume that the connections are equal to the number of households since most of the electricity consumers in rural areas are residential. Then the minimum population density in a province for an isolated grid in persons/km², $ND_{p,\min}$, can be determined as follows:

$$ND_{p,\min} = \frac{c}{10^{-6}\pi r^2} \times \overline{N}_{HH,p,rur} \qquad (5.16)$$

where $\overline{N}_{HH,p,rur}$ is the average household size in rural areas of province p.

With an average household size of about 4 persons in rural areas in Indonesia, the average minimum population density for a local grid in Indonesia is 525 persons/km².

The population living in areas with a population density larger than $ND_{p,\min}$ are assumed to be served with a local grid. This area, $A_{p,rur,lg,ND}$, is determined by:

$$A_{p,rur,lg,ND} = \frac{-1}{t_p} \ln\left(\frac{ND_{p,\min}}{ND_{rur,grid}}\right) \qquad (5.17)$$

Subsequently, the population living in this area, $N_{p,lg,ND}$, is obtained by:

$$N_{p,lg,ND} = \sum_{i=1}^{A_{p,rur,lg,ND}} ND_i \qquad (5.18)$$

where

$$ND_i = ND_{rur,grid}e^{-t_pi} \qquad (5.19)$$

Besides, if the population density is less than $ND_{p,min}$, different factors, e.g. the load size and willingness to pay, determine whether a local grid is viable or not. To meet the criteria for a local grid, the minimum load L should be roughly 10 kW [10]. When this criterion has been met, it is assumed that in 30% of these cases a local grid is feasible.

Furthermore, when a local grid is possible, the electricity demand of the settlement is considered as high, otherwise the settlement's electricity demand falls in the low demand category. Based on the minimum load requirement, the part of the rural population without electricity access which could be served by a local grid, $N_{p,lg,L}$, can be determined as follows:

$$N_{p,lg,L} = 0.3 \times \sum_{i=A_{p,rur,lg}}^{A_{p,rur,ne}} \left\{ ND_i | \frac{ND_i}{\overline{N}_{HH,p,rur}} L_{peak,h} > 10 \right\} \qquad (5.20)$$

The total population served by a local grid is, therefore:

$$N_{p,lg} = N_{p,lg,ND} + N_{p,lg,L} \qquad (5.21)$$

Subsequently, the population served by stand-alone PV systems, $N_{p,sa}$, will be:

$$N_{p,sa} = N_{p,rur,ne} - N_{p,lg} \qquad (5.22)$$

where $N_{p,rur,ne}$ is the total rural population in province p with no electricity access.

5.2.2.5 PV System Sizing

System sizing is based on the assumed electricity demand scenarios (Table 5.1) and the available irradiation based on data from NASA[3] which has been averaged over each $1 \times 1°$ latitude/longitude grid cell inside the bounding box area of each province, which is obtained from [11].

---

[3]Data based on monthly averaged values per latitude/longitude grid cell for a 22-year period (July 1983–June 2005).

The following formulas related to system sizing are presented for system configuration a. If not explicitly mentioned, the same formulas can be applied for configuration b, by replacing the subscripts a and h by b and l, respectively.

The daily electricity delivered by the PV system for a household, $E_{PV,HH,a}$ (kWh/day), is determined as follows:

$$E_{PV,HH,a} = \frac{\alpha_h E_{HH,h}}{365\, PR_a} \tag{5.23}$$

where PR_a is the performance ratio of the PV system in configuration a.

The peak power of the PV modules for a household in province p, $P_{PV,HH,a,p}$ (kW$_p$), is determined as follows:

$$P_{PV,HH,a,p} = \frac{E_{PV,HH,a}}{H_p} \tag{5.24}$$

where H_p is the daily average horizontal irradiation (kWh/m^2/day) in province p, which is numerically identical to the peak sun hours[4]. It is assumed that the yearly difference in horizontal versus tilted irradiation in Indonesia is negligible, due to its location close to the equator.

The inverter capacity for a household in province p, $C_{inv,HH,a,p}$ (kW), is chosen to be f_{inv} times the total peak power of the PV modules if this is larger than the peak load, otherwise the inverter capacity is chosen to be equal to the peak load:

$$C_{inv,HH,a,p} = \begin{cases} f_{inv}\, P_{PV,HH,a,p}, & f_{inv}\, P_{PV,HH,a,p} > L_{peak,h} \\ L_{peak,h}, & f_{inv}\, P_{PV,HH,a,p} \leq L_{peak,h} \end{cases} \tag{5.25}$$

The battery capacity for a household $C_{batt,HH,a}$ (kWh) is determined as follows:

$$C_{batt,HH,a} = \frac{E_{PV,HH,a}\beta_a}{\eta_{wires}\eta_{ctrl}\eta_{batt} DOD} \tag{5.26}$$

where β_a is the number of days of autonomy, η_{wires}, η_{ctrl} and η_{batt} are the efficiencies of the wires, charge controller and battery charge, respectively, and DOD is the battery's depth of discharge.

In the hybrid configuration, the diesel genset will mainly produce electricity during the peak hours in the evening. In this case, the genset can be sized to run efficiently during these peak hours, and the times of lower power demand during the day and night will be covered by the PV-battery system. The nominal power of the diesel genset for a household, $P_{DG,HH}$ (kW), is determined as follows:

$$P_{DG,HH} = \frac{L_{peak,h}}{DG_{BEP}} \tag{5.27}$$

---

[4]A peak sun hour is defined as 1 kW/m^2.

Table 5.4 Assumptions for system sizing

Variable	Symbol	Value	Unit
Diesel generator efficiency (a)	$\eta_{DG,a}$	30	%
Diesel generator efficiency (a_{ref})	$\eta_{DG,a,\text{ref}}$	25	%
Diesel generator efficiency (b_{ref})	$\eta_{DG,b,\text{ref}}$	20	%
Diesel generator best efficiency point	DG_{BEP}	80	%
Wires efficiency	η_{wires}	97	%
Battery charge controller efficiency	η_{ctrl}	98	%
Inverter efficiency	η_{inv}	95	%
Inverter capacity factor	f_{inv}	0.83	–
Battery (dis)charge efficiency	η_{batt}	95	%
Depth of discharge	DOD	80	%
Autonomy (a)	β_a	60	%
Autonomy (b)	β_b	200	%

Different values for the efficiency of diesel generators are given, corresponding to the system configurations (a) and (b). The subscript ref indicates the efficiency used in the reference scenario in which only diesel generators are used to supply electricity

where DG_{BEP} is the best efficiency point (BEP) of the diesel genset.

The assumed values used to size the PV systems, batteries, and diesel generators are presented in Table 5.4.

The battery self-discharge is assumed to be negligible due to the short-term storage. For standalone PV systems the performance ratio, PR_b, is assumed to be 55%, due to (i) mismatch between the time of electricity use and production, (ii) shadow losses and (iii) higher PV module temperatures in these tropical areas. For PV systems in an isolated grid a performance ratio, PR_a, of 70% is assumed.

5.2.3 Costs

The costs will be determined by calculating the levelized cost of energy (LCOE) according to:

$$\text{LCOE}_p = \frac{I_p + \sum_{n=1}^{L} \frac{AO_p}{(1+DR)^n} - \frac{RV_p}{(1+DR)^n}}{\sum_{n=1}^{N} \frac{E_{p,PV} \times (1-SDR)^n + E_{d,p}}{(1+DR)^n}} \tag{5.28}$$

where $LCOE_p$ is the total LCOE in province p, I_p is the initial investment for the PV system ($), AO is the annual operational costs ($), RV is the residual value ($ as % of investment), DR is the discount rate (%), $E_{p,PV}$ is the annual electricity production by PV (kWh/year), SDR is the system degradation rate (%), E_d is the electricity production by diesel gensets (kWh/year) and L is the project's lifetime in years.

For grid-connected PV the parameters will be calculated slightly differently compared with off-grid PV, therefore we divide the cost modeling steps into two sections.

5.2.3.1 Grid-Connected PV Systems

For grid-connected PV we start by calculating the nominal PV system capacity, C, in watt-peak (kW$_p$) per area type (urban, etc.) for each province:

$$C_{p,i} = \frac{E_{p,\text{PV},i} \times 10^6}{365 \times H_p \times PR_i} \tag{5.29}$$

where H_p is the daily average irradiation per square meter (kWh/m^2/d) and PR_i is the performance ratio of PV systems in area type i.

The initial investment, $I_{p,i}$, for each area type can be calculated as follows:

$$I_{p,i} = C_{p,i} \times P \tag{5.30}$$

where the installed system price (including BOS), P, is in \$/kW$_p$.

$$RV_{p,i} = I_{p,i} \times RV \tag{5.31}$$

where $RV_{p,i}$ is the residual value of the PV system in province p and area type i after its lifetime and RV is a percentage of the initial investment representing its salvage value.

The annual operation costs AO in \$/kW$_p$ are determined as follows:

$$AO_{p,i} = C_{p,i} \times AO \tag{5.32}$$

In Table 5.5 assumptions are shown for the calculation of the LCOE. The discount rate is based on the rate of the central bank of Indonesia [12]. A recent report by IRENA shows the cost breakdown for utility-scale PV systems' total installed costs in Indonesia. In total, the system price is estimated to be 1.192 USD$_{2018}$/kW$_p$. Compared

Table 5.5 Assumptions for LCOE calculation of grid-connected PV systems

Variable	Value	Unit
Lifetime (L)	25	years
Annual operations (AO)	15	\$/kW$_p$
Residual value (RV)	0.5	%
Discount rate (DR)	5.75	%
System degradation rate (SDR)	0.5	%
System price (P) [13]	1,192	\$/kW$_p$

with other G20 countries, Indonesia ranks relatively low (6th lowest), although the cost for PV modules (roughly 50% of the total installation costs) is relatively high (3rd highest, after Canada and South Africa). However, due to the lower margin and labor costs compared with other countries, the overall system price remains relatively low.

5.2.3.2   Off-Grid PV Systems

The costs of the PV system are included in the initial investment I_{PV}, while the costs of the other components (e.g. batteries, fuel) are included in the annual costs AC, which include maintenance and replacement costs as well. To determine the LCOE a number of cost assumptions have been made, see Table 5.6.

The cost of diesel is based on the fuel costs from PLN per province from 2018 [3] and has been converted to USD. In this study the exchange rate of 14,233.61 IDR/USD has been applied, which is the average for the year 2018, based on monthly values. Source: https://www.ofx.com/en-au/forex-news/historical-exchange-rates/monthly-average-rates/, resulting in an average cost of diesel in Indonesia of 0.64 $/l. The fuel costs from PLN already include transportation costs,

Table 5.6 Assumptions for the calculation of the LCOE of off-grid PV systems

Variable	Value	Unit
Investment costs		
Battery cost	150	$/kWh
Diesel generator	650	$/kW
O&M costs		
Annual operations PV	15	$/kW/year
Fixed O&M costs diesel	15	$/kW/year
Variable O&M costs diesel	0.03	$/kWh
Fuel		
Energy density diesel	35.86	MJ/l
Additional transport costs (*a*)	10	%
Additional transport costs (*b*)	50	%
Lifetime		
Diesel	20	years
PV (*N*)	25	years
Inverter	12	years
Battery	5	years
LCOE		
Residual value (*RV*)	0.5	% of $I_{PV,p}$
Discount rate (*DR*)	5.75	%
System degradation rate (*SDR*)	0.5	%

however these costs represent only the average fuel costs in the areas in which PLN operates. Since these are mainly urban areas, the fuel costs for rural areas would be higher, therefore extra costs are taken into account for the transportation of fuel, as shown in Table 5.6. For the hybrid system configuration *a*, the influence of transportation costs is assumed to be small since a micro-grid is only viable in more populated areas. In contrast, for the stand-alone PV system configuration *b*, the influence is assumed to be higher, since these off-grid systems are mainly located in sparsely populated remote areas. The additional modeled transportation costs are quite modest; in the study of Blum et al., factors of 2.0 and 2.73 are used for the medium and high-cost scenario, respectively [14].

5.3 Potential for Grid-Connected PV Systems

The potential of grid-connected PV systems in areas with a grid is estimated to be 3.18 TW_p, potentially generating 5.18 PWh annually, which corresponds to 18 times the total amount of electricity sold by PLN in 2018 (Fig. 5.4). The majority (82%) can be installed in rural areas; roughly 17% in suburban areas and the remaining approximately 2% in city centers. This opens an opportunity for Indonesia to export excess electricity to neighboring countries such as Malaysia in Kalimantan and the Malaysian Peninsula, Papua New Guinea in Papua, Timor Leste in Timor, and Singapore.

However, taking into account the limitation of the actual electricity demand during daylight time, the potential is estimated to be 73.2 GW_p potentially generating 95 TWh, which corresponds to 41% of the total amount of electricity sold by PLN in 2018 (Fig. 5.5).

In Fig. 5.4 the LCOE (in USD/MWh) and potential electricity generated by PV (in GWh/year) is shown for each province in Indonesia. The LCOE varies between 59 and 83 USD/MWh for suburban areas. For urban cores, the LCOE is in the range of 63–88 USD/MWh and for rural areas, the LCOE is estimated to be in between 67–95 USD/MWh. The differences in LCOE are due to the assumed performance ratios in the three different areas.

Based on PLN's overall generation costs, the average generating cost of PLN is roughly 54–63 USD/MWh, which is slightly lower than the LCOE of grid-connected PV. The average generation cost is highly influenced by the efficient large-scale electricity production on the islands Java and Bali. Therefore, in more remote places, the generation cost will probably be significantly higher. However, PLN has not published its generation costs per province. With an LCOE of roughly 60 USD/MWh, grid-connected PV should presumably be feasible in North Maluku, Maluku, and East Nusa Tenggara. Furthermore, it is assumed that outside the Java-Bali system, in many other places grid-connected PV is already cost-effective. At the same time, PLN's average selling prices vary in between 58 and 103 USD/MWh, depending on customer type [3]. For all provinces the selling price range is 26–123 USD/MWh

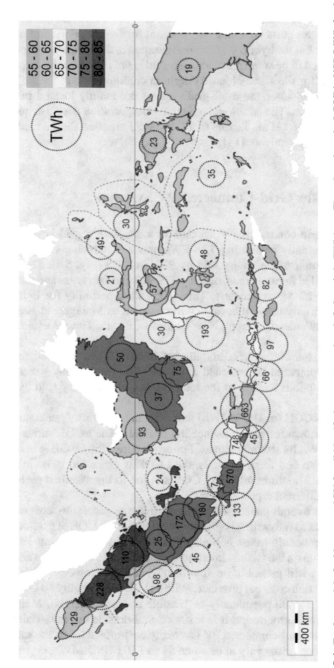

Fig. 5.4 Total potential of grid-connected PV for all identified areas per province in Indonesia in TWh/year. The colors show the LCOE in USD/MWh for the suburban areas

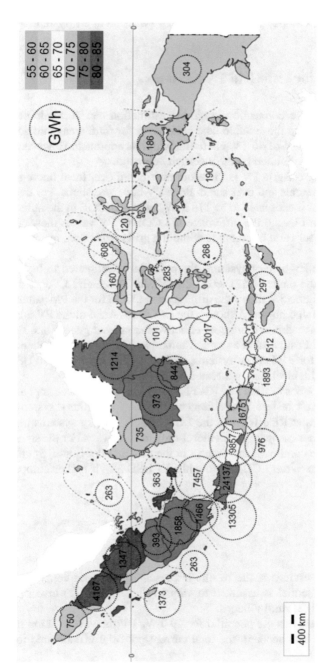

Fig. 5.5 Total potential of grid-connected PV for all identified areas limited by the actual electricity demand per province in Indonesia in GWh/year. The colors show the LCOE in USD/MWh for the suburban areas

[3]. Comparing these prices with the LCOE of grid-connected PV shows that part of PLN's customers could pay a lower price if they switched to PV systems.

5.4   Potential for Off-Grid PV Systems

For areas without grid connection, a simple estimation can be made of the off-grid potential by taking the suitable land area factor for rural areas into account. A potential for off-grid PV of 46 TW_p is found with this approach. However, here the actual potential is also limited by the electricity demand.

The potential for off-grid PV is therefore distributed over local microgrids consisting of hybrid systems and over stand-alone PV-battery systems. For microgrids, the overall potential is estimated to be 178 MW_p with an LCOE in the range of 211–320 USD/MWh. In Fig. 5.6 the difference in LCOE of PV versus diesel is shown. In every province, for rural households without a grid connection, microgrids show a lower LCOE.

For stand-alone PV-battery systems, the potential is estimated to be 168 MW_p with an LCOE in the range of 571–605 USD/MWh. This high LCOE is due to the cost of batteries. Figure 5.7 shows the difference in LCOE for the PV-battery system versus diesel-generated power. Although the LCOE for stand-alone PV is relatively high, in most provinces it is more cost-effective than diesel generators. Especially in the provinces of East Nusa Tenggara, Maluku, and North Maluku the conditions are very favorable for PV-battery systems, with an LCOE more than 30 USD/MWh lower compared with electricity generated by diesel generators.

An exception applies to Riau and Riau Islands, where fuel prices kept artificially low at 0.17 $/l result in less favorable conditions for PV-battery systems. In the provinces of Riau and Riau Islands, the LCOE of PV-battery systems are 27 and 26 USD/MWh higher compared with diesel-only, respectively. For the same reason, similar unfavorable conditions are present in the provinces of South, Southeast and West Sulawesi, where the LCOE is roughly 15 USD/MWh higher compared to a diesel-only system.

5.5   Overview

In Table 5.7 an overview of the results is presented. For grid-connected PV the demand-limited potential is assumed to vary in the order of (1) urban cores, (2) suburban areas and (3) rural villages.

In total, Indonesia has the potential for 49 TW_p PV; however, taking the actual electricity demand into account, the total current potential is estimated to be 73.5 GW_p.

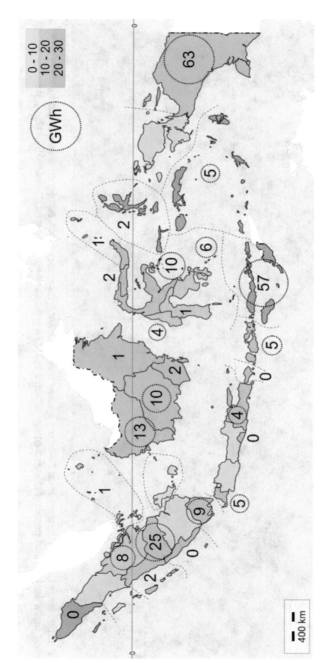

Fig. 5.6 Difference in LCOE of the hybrid PV-battery-diesel configuration versus diesel-only for each province. Positive differences indicate a lower LCOE for PV. The dashed circles show the total amount of electrical energy delivered by PV. Grey provinces do not have rural households lacking grid access

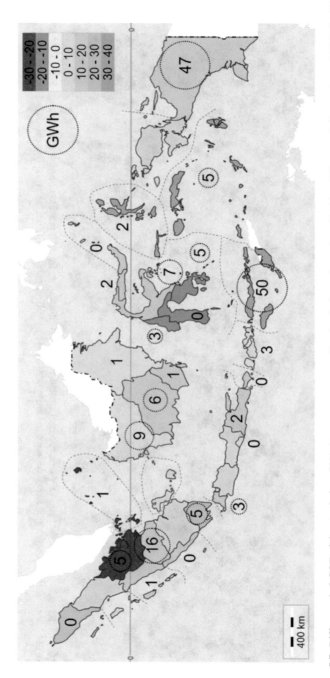

Fig. 5.7 Difference in LCOE of the stand-alone PV-battery systems versus diesel-only for each province. Positive differences indicate a lower LCOE for PV. The dashed circles show the total amount of electrical energy delivered by PV. Grey provinces do not have rural households lacking grid access

Table 5.7 Overview of the results for grid-connected and off-grid PV systems based on the available area and limited by demand

Category	Type	Potential (GWp)		Average LCOE (USD/MWh)
		Suitable area	Demand limited	
Grid-connected PV	Urban cores	56	48.8	78
	Suburbs	533	24.4	73
	Rural villages	2,595	0.1	83
Off-grid PV	Hybrid	45,949	0.2	281
	Stand-alone	n.a.	0.2	590
Total		49,132	73.5	

5.6 Discussion

This study estimates the technical potential and cost (LCOE) of PV systems in Indonesia. Many assumptions have been made and the model is a simplification of reality, but from this high-level study, it becomes clear that the potential for installation of PV systems in Indonesia is enormous and that PV systems, both grid-connected as off-grid systems, are already affordable in many areas.

In this study the estimated LCOE varies between 59 and 95 USD/MWh for grid-connected PV systems in Indonesia. According to the report from IRENA [13], the global weighted average LCOE of utility-scale PV system projects in 2018 has been found to be 85 USD/MWh. The values in this study are in line with these global values. Furthermore, the same report shows that the LCOE of utility-scale PV systems is expected to fall to 48 USD/MWh in 2020.

The LCOE of stand-alone PV systems is still relatively high, because of the assumption of a lower PR for the PV system and the battery costs. The latter is influenced by the assumed energy demand and days of autonomy. Therefore, energy savings, for instance by using energy-efficient electrical appliances, will reduce the energy demand and as such the overall LCOE of stand-alone PV systems.

With further decreases in the costs of PV and battery technologies, the conditions for PV will become even more favorable, resulting in more areas where PV systems will be a financially feasible alternative.

The Indonesian government aims to reduce 29% of its GHG emissions against the business-as-usual scenario by 2030. In order to do so, the government has various options. Based on a study by Handayani et al. [15] the most cost-effective way for capacity expansion of the Java-Bali system while meeting the climate goals is by increasing renewable energy development in combination with natural gas. From this study, it is found that outside the Java-Bali system, generally PV systems have a lower LCOE than electricity from other sources. Therefore, it would make sense to stimulate grid-connected PV in the provinces outside Java-Bali; in particular, off-grid PV systems can play a large role in the electrification of remote areas.

A recent study on renewable energy policy in Indonesia emphasizes the importance of engaging the private sector in order to stimulate growth in the installed clean energy capacity [16]. A report from PricewaterhouseCoopers (PWC) presenting results of a survey in the power industry in Indonesia shows, however, that regulatory changes to the Independent Power Producer (IPP) investment framework by the government will not help to create certainty for private companies to invest in the power sector [17]. It is, therefore, necessary to develop long-term policies which enable low-risk investments by the private sector in order to achieve the electricity expansion and climate goals.

5.7   Conclusions

The potential of grid-connected PV systems in areas with a grid is estimated to be 3.18 TW_p, potentially generating 5.18 PWh annually, which corresponds to 18 times the total amount of electricity sold by PLN in 2018. The majority (82%) can be installed in rural areas; roughly 17% in suburban areas and the remaining approximately 2% in city centers.

Taking into account the limitation of the actual electricity demand during daylight time, the potential is estimated to be 73.2 GW_p potentially generating 95 TWh, which corresponds to 41% of the total amount of electricity sold by PLN in 2018. The LCOE of grid-connected PV varies between 59 and 95 USD/MWh, depending on the location.

For off-grid PV systems, the potential is roughly 46 TW_p. By taking the actual electricity demand of rural households which lack a grid connection into account, the total potential is found to be 178 and 168 MW_p for hybrid and stand-alone PV systems, respectively. Hybrid systems have an average LCOE of 281 USD/MWh; stand-alone systems have an average LCOE of 590 USD/MWh.

In total, Indonesia could host 49 TW_p of PV systems; however, taking the actual electricity demand into account, the total potential for PV systems is estimated to be 734 GW_p (in 2018).

Finally, it is important to emphasize that both grid-connected PV systems and off-grid PV systems are already financially feasible in large parts of Indonesia.

References

1. A.J. Veldhuis, A.H.M.E. Reinders, Reviewing the potential and cost-effectiveness of grid-connected solar PV in Indonesia on a provincial level. Renew. Sustain. Energy Rev. **27**, 315–324 (2013)
2. A.J. Veldhuis, A.H.M.E. Reinders, Reviewing the potential and cost-effectiveness of off-grid PV systems in Indonesia on a provincial level. Renew. Sustain. Energy Rev. **52**, 757–769 (2015)
3. PT PLN (Persero), *Statistics PLN 2018* (2019)
4. Badan Pusat Statistik, *2010 Population Census* (2010)

5. Statistics Indonesia, *Population Projection by Province, 2010–2035* (2014)
6. Statistics Indonesia, *Percentage of Urban Population by Province, 2010–2035* (2014)
7. Badan Pusat Statistik, *Household by Region and Primary Source of Lighting in Dwelling Unit* (2010)
8. IEG, The welfare impacts of rural electrification: a reassessment of the costs and benefits. An IEG impact evaluation (Independent Evaluation Group, World Bank Washington^ eD. C D. C, 2008)
9. UNFCCC, Annex 5—Rationale for default factors used in the proposed methodology SSC-I.L Electrifi. Rural Communities Renew. Energy (2012)
10. SWECO, Assessing technology options for rural electrification. Guidelines for project development (2009)
11. NASA, *Atmospheric Science Data Center* (2005)
12. Trading Economics, *Indonesia Interest Rate* (2019)
13. IRENA, *Renewable Power Generation Costs in 2018* (Abu Dhabi, 2019)
14. N.U. Blum, R. Sryantoro Wakeling, T.S. Schmidt, Rural electrification through village grids—Assessing the cost competitiveness of isolated renewable energy technologies in Indonesia. Renew. Sustain. Energy Rev. **22**, 482–496
15. K. Handayani, Y. Krozer, T. Filatova, Trade-offs between electrification and climate change mitigation: an analysis of the Java-Bali power system in Indonesia. Appl. Energy **208**, 1020–1037 (2017)
16. M. Maulidia, P. Dargusch, P. Ashworth, F. Ardiansyah, Rethinking renewable energy targets and electricity sector reform in Indonesia: a private sector perspective. Renew. Sustain. Energy Rev. **101**, 231–247 (2019)
17. PWC, *Alternating Currents: Indonesian Power Industry Survey 2018* (2018)

Appendix

A.1 Demographics of Respondents Who Participated in the User Study

See Table A.1.

Table A.1 Demographics of respondents who participated in the user study

No. of respondents	Pekanbaru	Kupang	Jayapura
	114	65	26
The distribution of respondentsby city address (%)			
– Urban-core	53	45	23
– Sub-urban	47	55	77
Monthly Income(%)			
– High	28	5	4
– Upper-Middle	43	28	50
– Lower-Middle	17	49	27
– Low	6	15	0
– Not answer	6	3	19
Sex (%)			
– Male	49	51	42
– Female	51	49	58
Age groups (%)			
– 18–29	11	5	31
– 30–49	63	54	58

(continued)

K. Kunaifi et al., *The Electricity Grid in Indonesia*,
SpringerBriefs in Applied Sciences and Technology,
https://doi.org/10.1007/978-3-030-38342-8

Table A.1 (continued)

No. of respondents	Pekanbaru	Kupang	Jayapura
	114	65	26
– 50–64	23	35	12
– 65+	4	6	0
Education (%)			
– No school	0	2	0
– Basic school	8	14	0
– High school	50	42	0
– Undergraduate	30	38	8
– Postgraduate	5	3	85
– No answer	7	2	8

A.2   The Survey Sheet (English-Translated)

A.   Respondent Information

Income group: ET / EM / ES / ER (to be filled by the surveyor)
Sex   : M / F (circle an answer)
Age   :
Occupation   : _____
Education   : No formal school /Primary / Junior H. / Senior H. / Diploma / UG /
                  M / PhD *(circle an answer)*
Marital status : Not married / Married / Widow (circle an answer)
Ethnic group : _____
Address   :

Confidentiality
Your personal information will be kept confidential and be treated based on
standard practice. You can also mention below if there are other information you
will provide that you would like to be confidential as well.

_____

Signature
_____

B.   Questions

Please circle (O) on the most proper answers.

1.   Would you accept an increase in your electricity bill for better electricity service?

    a. Yes

    b. No

2. How much increase in your electricity bill would you find acceptable?

    a. 10–30%

    b. 30–50%

    c. 50–70%

3. Do you have a backup generator at home?

    a. Yes

    b. No

4. Do you experience a stable electricity voltage at home?

    a. Yes

    b. No

5. Have you ever experienced a blackout at home?

    a. Yes

    b. No

6. On average, how often in a month do you experience blackouts?

    a. less than 3 x

    b. 3–5 x

    c. 6–10 x

    d. more than 10 x

7. On average, how long is the duration of the blackouts you experience?

    a. <5 min

    b. 5–15 min

    c. 15–60 min

    d. >120 min

8. At what time of day would a blackout event incur the most losses for you?

    a. 6 a.m.–12 a.m.

    b. 12 a.m.–6 p.m.

    c. 6 p.m.–12 p.m.

    d. 12 p.m.–6 a.m.

9. On average, what is the duration of a blackout that would incur economic losses for you?

    a. less than 5 min

    b. 5–15 min

c. 15–60 min
d. more than 120 min

A.3 The Proportion of Respondents' Answers to Survey Questions, in Percentages

See Table A.2.

A.4 The Proportion of Respondents' Answers to Survey Questions, in Percentages

See Table A.3.

B.1 Input Data Per Province for the Calculation of the Technical Potential for Grid-Connected PV Systems

Province	A ($\times$ 1000) (km^2)	N ($\times$ mln.)	UR (%)	N$_{urban\_core}$ ($\times$ 1000)	ER (%)	H (kWh/m^2/day)
Aceh	58.0	5.3	33.2	614	99.8	5.1
Bali	5.8	4.3	70.2	884	99.6	5.3
Bangka Belitung	16.4	1.4	56.0	201	100.0	4.5
Banten	9.7	12.6	69.9	4.8	93.6	4.8
Bengkulu	19.9	2.0	32.6	348	99.8	4.8
Central Java	32.8	34.5	51.3	3	100.0	5.5
Central Kalimantan	153.6	2.6	40.2	250	76.4	4.8
Central Sulawesi	61.8	3.0	30.5	363	83.0	5.0
East Java	47.8	39.6	54.7	51	94.9	4.9
East Kalimantan	204.5	4.3	68.9	2	97.0	4.8
East Nusa Tenggara	48.7	5.4	24.3	372	56.3	6.2

(continued)

(continued)

Province	A (× 1000) (km^2)	N (× mln.)	UR (%)	N$_{urban\_core}$ (× 1000)	ER (%)	H (kWh/m^2/day)
Gorontalo	11.3	1.2	44.0	203	89.8	5.1
Jakarta	0.7	10.5	100.0	10.6	100.0	4.8
Jambi	50.1	3.5	33.3	674	51.9	4.6
Lampung	34.6	8.4	31.3	1.1	92.7	4.9
Maluku	46.9	1.8	38.9	397	85.8	5.8
North Maluku	32.0	1.2	28.5	236	87.2	6.0
North Sulawesi	13.9	2.5	54.7	756	97.3	5.9
North Sumatra	73.0	14.6	56.3	3.5	100.0	4.5
Papua	319.0	3.3	31.2	281	42.5	5.0
Riau	87.0	6.8	40.1	1.4	89.1	4.4
Riau Islands	8.2	0.8	83.3	1.5	84.7	5.1
South Kalimantan	38.7	4.2	48.4	951	96.5	4.8
South Sulawesi	46.7	8.8	45.0	1.7	99.5	5.4
South Sumatra	91.6	8.4	37.3	2.1	100.0	4.6
South-East Sulawesi	38.1	2.6	35.0	456	84.9	4.9
West Java	35.4	48.7	78.7	10.7	100.0	4.8
West Kalimantan	147.3	5.0	36.2	795	87.0	5.0
West Nusa Tenggara	18.6	5.1	49.4	584	90.4	5.6
West Papua	97.0	1.0	34.9	151	100.0	5.1
West Sulawesi	16.8	1.3	23.0	–	78.4	5.5
West Sumatra	42.0	5.4	49.6	1.4	97.4	4.9
Yogyakarta	3.1	3.8	74.6	435	94.9	4.8

Table A.2 The proportion of respondents' answers to survey questions, in percentages

Question 1	Would you accept an increase in your electricity bill for better electricity service?								
Answer	Yes			No			NA		
	P	K	J	P	K	J	P	K	J
	56	68	58	35	31	38	9	2	4

Question 2	How much increase in your electricity bill would you find acceptable?											
Answer	10–30%			30–50%			50–70%			NA		
	P	K	J	P	K	J	P	K	J	P	K	J
	89	55	50	7	0	0	0	0	0	4	45	50

Question 3	Do you have a backup generator at home?								
Answer	Yes			No			NA		
	P	K	J	P	K	J	P	K	J
	21	14	65	67	83	27	12	3	8

Question 4	Do you experience a stable electricity voltage at home?								
Answer	Yes			No			NA		
	P	K	J	P	K	J	P	K	J
	82	49	50	11	48	42	8	3	8

Question 5	Have you ever experienced a blackout at home?								
Answer	Yes			No			NA		
	P	K	J	P	K	J	P	K	J
	89	97	85	2	0	8	10	3	8

Question 6	On average, how often in a month do you experience blackouts?				
Answer	<3 x	3–5 x	6–10	>10 x	NA

(continued)

Table A.2 (continued)

	P	K	J	P	K	J	P	K	J	P	K	J	P	K	J
	37	37	8	39	25	46	10	31	35	10	5	4	4	3	8

Question 7 — On average, how long (in minutes) is the duration of the blackouts you experience?

Answer	<5			5–15			15–60			60–120			>120			NA		
	P	**K**	J	P	K	J	P	K	J	P	K	J	P	K	J	P	K	J
	0	**0**	0	3	5	12	19	12	15	60	66	54	14	12	12	4	5	8

Question 8 — At what time of day would a blackout event incur the most losses for you?

Answer	6 a.m.–12 a.m.			12 a.m.–6 p.m.			6 p.m.–12 p.m.			12 p.m.–6 a.m.			NA		
	P	K	J	P	K	J	P	K	J	P	K	J	P	K	J
	27	52	58	13	8	4	52	29	23	2	0	0	6	11	15

Question 9 — On average, what is the duration (in minutes) of a blackout that would incur economic losses for you?

Answer	<5			5–15			15–60			60–120			>120			NA		
	P	K	J	P	K	J	P	K	J	P	K	J	P	K	J	P	K	J
	3	2	0	2	2	8	6	12	12	72	72	54	4	6	12	13	6	12

P Pekanbaru, *K* Kupang, *J* Jayapura, *NA* No Answer

Table A.3 The proportion of respondents' answers to survey questions, in percentages

Question	P	K	J*	P	K	J*	P	K	J*	P	K (Don't know)	J*	P	K (NA)	J*
Have you heard of 'renewable energy'?	34	No 37	19	55	Yes 60	77							12	2	1
Is renewable energy important for Indonesia?	51	Important 57	81	3	Neutral 2	15	1	Not important –	–	4	5	1	48	24	1
Have you heard of 'climate change'?	69	Yes 58	92	20	No 38	4							11	2	1
Are you worried about climate change?	60	Yes 51	85	7	Neutral 5	8	1	No 3	–	2	3	2	35	25	2
Have you heard of 'PV systems'?	25	No 38	4	65	Yes 58	88							11	2	2
Which one of the following two electricity sources you believe is cheaper?	17	PLN –	19	29	PV 49	54				45	46	23	11	NA	1
Which one of the following two electricity sources you believe is better for the environment?	9	PLN 5	8	42	PV 51	85				39	42	–	11	NA	2

(continued)

Table A.3 (continued)

	P	K	J*	P	K	J*	P	K	J*	P	K	J*	P	K	J*
Which one of the following two electricity sources you believe is more stable?		PLN			PV			Don't know			NA				
	25	15	31	15	31	19	50	49	42	11	3	2			
Which one of the following two electricity sources would you choose to provide power supply to your home?		PLN			PV on-grid			PV off-grid			Don't know			NA	
	30	15	19	30	17	46	8	25	12	23	35	15	11	5	2
Would you like that to have a PV system being installed on your house's rooftop?		Yes			No			NA							
	65	71	88	32	14	4	4	10	2						

Index

A

Acid rain, 4

Agriculture, 3, 5, 79

Alternating-Current (AC), 10, 14, 92

Archipelago, 1, 3, 4, 39, 41

Array, 8, 10, 70, 72

Association of Rooftop PV System Users (PPLSA), 17

Association of South-East Asian Nations (ASEAN), 5, 7, 35

Attitudes, 67, 68, 70, 73

Australia, 3, 5, 10, 12, 25

B

Back-up power, v

Bali, 3, 12, 14, 23, 25, 27, 31, 35, 39, 48, 93, 99, 106

Battery, 76, 82, 87, 89, 90, 92, 96–99

Battery charge, 89, 90

Best Efficiency Point (BEP), 90

Blackouts, 27, 37, 45, 46, 53, 54, 56, 57, 105, 108, 109

Borneo, 3

BOS, 91

Brunei Darussalam, 5

Build-Own-Operate-Transfer (BOOT), 32

Bureau of Meteorology, 47

Business, 7, 18, 21, 25, 26, 28, 30, 33–35, 54, 81, 99

Business models, 30

C

Cadmium telluride, 8

Carbon dioxide (CO_2), 28

Central government, 1, 14, 18

Charge controller, 89

Climate
change, 7, 21, 28, 35, 67–69, 73, 110

Coal, 21, 22, 28

Coastline, 3

Copper indium gallium selenide, 8

Correlation, 5, 6, 58–60

Costs, 4, 26, 31, 32, 34, 41, 43, 44, 51, 59, 61, 70, 75–77, 82, 83, 90–93, 96, 99

D

Days of autonomy, 89, 99

Degradation, 6, 7, 90–92

Demand, 1, 7, 21, 23, 25, 33, 37, 39, 76, 80–83, 86–89, 93, 95, 96, 99, 100

Depth Of Discharge (DOD), 89, 90

Diesel
fuel, 4
generator, 4, 26, 27, 76, 82, 87, 90, 92, 96

Direct Current (DC), 8, 10

Disasters, 3

Distribution Network (DN), 28, 29, 38, 41, 43, 44, 61

Droughts, 3

E

Earthquakes, 3

Eastern, 3, 4, 8, 14, 18, 27, 35, 39, 44

Economic
growth, 5–7, 22, 39

Printed in the United States
By Bookmasters